A Patient's Point of View

Hidden Abundance of Health to Expand Your Capacity to Heal

Matthew Blake

ISBN: 979-8-218-58324-8

Published by:
Harper Book Publisher
www.harperbookpublisher.com

Printed in the United States of America

Disclaimer:

This is my story of a path of healing and restoring health from several health problems that arose in my life. The information in this book is not to be taken as medical advice for your situation.

Dedication

This book is dedicated to

My lovely wife, Loren, who stood by my side and gave me tough love to heal while also holding our life and family together as I have been recovering to better health.

My loving mother, who, through our talks and visits, gave me the inspiration to heal and to know the best is yet to come.

My dearly missed father, whose memory and words to stay strong even when so much seems to go wrong and not in the right direction, gave me strength to continue onwards.

My incredible sister, without whose unwavering support and encouragement this book would not be possible.

My daughter, Molly, who wrote Psalms from the Bible on the white board in my hospital room, inspired me to heal and took the time to take care of me while in the hospital along with her husband, Bradley.

My daughter, Georgia, who kept my business clients and contacts updated during the initial hospital stay.

My youngest daughter, Lucy, who lovingly made a radiation sessions chain that kept me and our family motivated throughout.

My son, Jack, for a very special gift from him that extended my life and my quality of living.

Each of my children, Molly, Jack, Georgia, and Lucy, that were a constant motivation for me to get better and remain present in each of their lives.

My supportive brother, Michael, whose encouragement while also being my personal trainer to get me strong, added to my belief that I can accomplish anything.

Loren's parents, George and Kathy, who helped with things around the house and helped keep things going along by supporting Loren.

My doctors, therapists, and nurses who took great care of me.

Jesus Christ, my Lord and Savior.

All of the other people who helped me financially during my recovery.

Everyone who sent positive thoughts and/or prayers for me and for my health to be restored.

About the Author

Matthew Blake is a business trainer dedicated to the success of his clients and is the author of *"A Patient's Point of View: Abundance of Hidden Health That Expands Your Capacity to Heal."*

Matthew is an entrepreneur, husband, father, and son whose life catapulted him into a first-time author from life's unpredictable onset and resolution of a chronic affliction to the upheaval of sudden health challenges. Matthew shares his distinctive voice and experience of cultivating and maintaining resilience using an unwavering determination to triumphantly navigate through obstacles to regain, restore, and open new opportunities for physical, mental, and spiritual health.

Matthew's passion is to help others know their potential for success and abundance and to be a voice of hope in the world. He enjoys running and working out in a gym. Matthew is a big college football fan, and one of his favorite things to do on Saturday afternoons is to watch the University of Notre Dame play football.

Table of Contents

Introduction

Good health. We wish for good health, and we always want to have good health. When our health is good, our lives are also much better and greatly more successful in numerous ways. Yet, good health in us, as human beings is not an ironclad guarantee. This body of ours is sensitive, can be somewhat easily damaged, and gets off balance, which negatively affects the state of our good health. On the change of a dime, our relatively good health can suddenly topple into ill health and/or disease. This disruption within the lifeforce of good health now dictates an imposed course into unnavigated and unknown territory.

Novel chaos and disorder are now erupting into our carefully planned days and life goals. Disarray is ejecting into our daily lives and activities, ungrounding and unrooting our senses of security and ability to predict our lives and forecast our future. Such instability can leave us feeling robbed of what we cherish as important in our life, leaving us with feelings of confusion and despair and asking the question, "Why me? Why now?" All of this is the frame of mind that is scarce, a shortage, a lack of, and a deficit with impending misfortune.

One of the benefits of living in our world is that we live within the polarity of opposites. We have day and night, we have sun and moon, we have up and down, right and left. In whatever we think of, we can state the opposite of what we are thinking. You may ask, "How can this be a benefit?" Since we have opposite polarities, we are not absolutely bound to experience something in only one way. Change is ever possible; our mind just needs to be open to and then know this possibility. Therefore, not only does a mindset of scarcity, lack of, or not enough exist. Simultaneously existing is the mindset of boundless abundance.

This book is a real-life story about the onslaught of numerous threats that confronted and were hazards to the good health of the main protagonist, Matthew. When reading this book, clearly, the menaces of the threats to his health could have turned the tide to devastating outcomes, including death. In spite of this, Matthew rose above these perils utilizing what he terms the abundance mindset. He shares his insights with his "secret sauces" on how anyone can use an abundance mindset to not just overcome but even prosper beyond expectations of ourselves and others. Besides reading his truly amazing story, this book, filled with teachable moments, helpful pointers, valuable reminders, and how to directly access your hidden abundance to heal, is a must-read.

Preface

I had survived and gotten through the life-saving emergency procedure to remove the blood clot in my brain. After a few days' stint in the ICU, I was laying in bed in my regular hospital room, waiting for the nurse to escort me to the bathroom. I was not allowed to get out of the hospital bed without supervision. Earlier in the ICU, I was thinking about the homeopathic remedies I had been taking with positive results. I wasn't comfortable letting the staff know about the homeopathic remedies that I was taking prior to my unexpected visit to this hospital.

These thoughts were running through my mind as I lay in the hospital bed waiting for the nurse after ringing the call bell. I started thinking more about what it has been like being a patient in this situation, something I never expected to happen to me. What was being a patient like? How was being a patient affecting me? How did others like medical staff, friends, or family around me see me as being a patient? A thought came to my mind that patients do have a point of view on what is going on with them and around them. However, my feeling and experience was that often, the actual patient's point of view was not really understood, listened to, or incorporated into medical care and aspects of life beyond medical care. My experience was that there were many times when, rather than being talked to, I was being talked at. I thought it was not possible this had only happened to me.

As my thoughts delved more deeply into this, an idea came to my mind that I should write a book called "A Patient's Point of View." I wanted the world to know what a patient feels during a hospital stay and afterward when reintegrating back into living daily life, as well as what a patient goes through. I desired to give a voice to myself and others who have been living their life as a patient. I started thinking that writing a book could help me get on a platform nationally where I could help people train their minds to have an attitude of abundance no matter what happens. Maybe, besides writing a book, there would be opportunities to travel to do presentations and other ways to meet people. I desired to talk about this abundance of health we usually do not know is around us and within us.

A strong motivation arose within me to share with the world about the view of a patient receiving health care, and I wanted to show the possibilities of health when holding onto an attitude of abundance, even in very dark times. While still waiting for the nurse, I picked up my cell phone and texted my sister, *"Hi there. I want to write a book about being a patient called "A Patient's Point of View,"* which talks about healing and health. Would you be a part of the book and help write about some areas of the book?"

My sister, waiting in line at the post office to mail a package to me, had been thinking about me when my text came through. Immediately, she texted back, *"Wonderful! That's a great idea! Thanks for thinking of me to help you with some areas of the book. Looking forward to it! Do this! You can do this when you are ready!"* Getting that text at that moment meant so much to me. In the condition I was presently in, she just said, "Do this," and it was all I needed to hear. I spent time pulling myself and my life together again, step by step, moment by moment, after which an unexpected road appeared that I suddenly was traveling, whether I wanted to be on it or not. There was no exit ramp. Then, the time arrived, it was time to write this book, and here it is. With deep wishes, may this book offer you some sort of inspiration, the ability to expand your abundance and knowledge of how to shift your life into more bountiful healing and health that benefits your life and the people around you.

Chapter 1: Ignorance is Bliss - the First Encounter

The year was 1998. My beautiful wife, Loren, of two years, was six months pregnant with our first daughter when, out of the blue, I urinated blood. At that time, my wife and I were both in very good physical shape. We met at a fitness center where she was an aerobics instructor, and I was a personal trainer. Therefore, when this episode occurred, I was not very worried about it because I was in good physical shape. However, the reality of it was that I didn't want to know. Ignorance was bliss for me.

At the request of my wife, I went to see my primary care physician, and truth be told; I didn't have a great opinion of doctors. This was because of my experience with my primary care physician that even though he was a nice guy, he pushed medication on me for simple things. Begrudgingly, I went, and he did a urinalysis test on me to find out that protein was in my urine. He suggested that I go see a nephrologist, which is a kidney doctor.

This is where the journey begins. I found a nephrologist and went to see him. He was a nice guy, and he recommended a kidney biopsy. So, I did get a kidney biopsy, and this is when things didn't start well. As a result, I started not trusting doctors. When I went to see the nephrologist after the biopsy, he literally read my results in front of me. In other words, he had not reviewed my medical chart or done any research prior to my appointment. That really turned me off to his suggestion, after skimming through my records, that I go on Prednisone, a steroid used for kidney disease. He stated that I had a progressive kidney disease called Focal Glomerular Sclerosis (FGS), and Prednisone had only a 30% chance of slowing down the disease. However, this was the "accepted medical treatment," so I should start right away.

FGS is a disease in which scar tissue develops in an area of the kidney called the Glomeruli, which are tiny parts of the kidney responsible for filtering waste from the blood. In my situation, it is not known why this started in the kidneys. Over time, the kidneys will fail with the build-up of so much scar tissue, and the only option to treat this is a kidney transplant, or the kidneys will fail completely, resulting in death.

I remember his response to me like it was yesterday: "Mr. Blake, your wife seems to be taking this news much better than you are." My response was, "Doctor, with all due respect, you flipped through my chart in a couple of minutes, and you just said to me that I have a progressive disease and I have to start Prednisone immediately, which has a lot of side effects and has a 70% chance of not working. So, forgive me if I'm not taking this news well."

My wife and I left that meeting and decided to go get a second opinion. We searched out numerous nephrologists, and we settled on a one-on-one consult at Johns Hopkins University in Baltimore, Maryland. The big reason why I decided on the nephrologist at Johns Hopkins University Hospital was because he wasn't opposed to alternative medicine. He didn't think it would work, but he wasn't opposed to it, which made me feel comfortable. This turned out to be a 20-year relationship.

This progressive kidney disease lasted 23 years before needing any medical intervention, far beyond any expectations that the medical community thought was possible. Through the course of this time, I did use some alternative health methods like acupuncture, which was actually very helpful and, I fully believe, slowed the deterioration of the kidneys. It is very rare to be able to live with this progressive kidney disease for so many years. This was such a great opportunity, and it also gave me false hope that I could handle anything.

At the time when my journey with this progressive kidney disease FGS started, although it was difficult to deal with, I was 28 years old and in good physical shape. Even though it was suggested by the nephrologist to go on blood pressure medication due to the kidney problem, it just didn't make sense to me at this time. Little did I know that I was spiritually being protected.

I was so self-centered on what I could do as opposed to what I could not do. I was focused on how to heal. For the next 23 years, I lived with this kidney disease while in the back of my mind, I really was hoping that all of this would just disappear. Every time I went to see my nephrologist, it was very contentious because I asked a lot of questions. I didn't just take what he said to do because I was very skeptical of the medication that I would have to take. I felt that the "cure" with the medication could be worse than the actual disease. Looking at all of this just from a worldly view of sickness and health just didn't work for me. I say all this because there's real hope in the world; we just need to be open to this knowledge. For me, this has been through a supernatural and spiritual view that blessings in my life occur through God and my Lord and Savior, Jesus Christ. For others, seeing the hope in their world may come through other spiritual paths or various avenues. The feeling and reassurance of hope within this world deeply assisted me in traveling this road with this kidney disease.

At the time all this started, I was young, ambitious, and very hardheaded. I believed that I could handle whatever was thrown at me. Even though I was following my own spiritual path of following Jesus' teachings, looking back, I felt that I never really thanked him or God for helping me have 20 years of relatively good health.

From my personal experience, I feel that it was not just alternative health methods, medical follow-ups, and a good diet for kidneys that helped me maintain relatively good health all of that time. I also feel that there was spiritual energy supporting me. During these 23 years, all that I really had was high blood pressure and bouts of gout due to high uric acid that occurs from damage to the kidneys' filtration system, so they can't filter waste products very effectively or efficiently.

Everything caught up to me very suddenly during the unprecedented pandemic year of 2020, when my kidneys began to fail. My blood pressure was going through the roof, and in every bloodwork test, it came back that my kidney function was declining quickly. So, my answer was to exercise harder, and all this would go away, but it didn't. The COVID-19 pandemic was frightening in the hospitals, and I took a bigger interest in the pandemic than in my own personal health.

However, the day came when, again, my guardian angel, my wife, took my blood pressure, and it was so high and alarming she immediately put me in the car and took me to the hospital. While in the hospital, I thought I would get some new blood pressure medication, and then all this would be under control. Well, this did not happen. When the doctor saw me, he wanted to admit me. Some bloodwork that was done found out that due to the kidney disease, my kidneys were not functioning well. This was like a gut punch for me because, previously, I was always able to bounce back. I was admitted to the hospital, which was difficult because of the pandemic. I was very alone because I could only have one visitor a day.

Then, I got the news from the nephrologist that it was time to go on dialysis. My first reaction was very skeptical because I thought I could just beat this. This doctor said to me that if I was going to resist doing dialysis, it was going to be the end of my life. I realized I had to do this, which I did agree to do. To complicate matters, I was put on a bunch of new blood pressure medications.

After a few days, I was discharged from the hospital and returned to my home. On my first day at home, I was supposed to take the medication on the same schedule that I was on in the hospital. In the hospital, I was getting the medication early in the morning and then again at lunch. When I got up at home in the morning, it was later than when I was in the hospital. I took the medication in the late morning. About one hour later, it was lunchtime, and I retook the medication since this is what I had done in the hospital.

Unknowingly, I took the medication too close together and did not realize that this could be a serious problem. Unfortunately, this resulted in my first stroke. Reflecting back, I realize I was having a stroke because I was dizzy and my words were slurred. However, at the time, I thought I was just reacting to the medication. I laid down on the couch to rest for the entire rest of the day. A theme you're going to hear in this book is how often I felt helped spiritually and protected because I spent 24 hours at home before going to the hospital after this happened.

When we arrived at the hospital, everybody was wearing a mask, including me, because of the pandemic. I could not explain why I was back at the hospital to the people who were trying to admit me. The nurses and the doctors quickly did a CT scan and realized that I had a stroke. I was completely shocked and scared to hear, "You've had a stroke." In my spiritual view, by the grace of God, there were no residual effects from the stroke. Incredibly, my speech cleared up, and I was able to function as if a stroke had never happened. I was able to function normally. My focus was to be discharged from the hospital and get back to my normal life.

My normal life during this time was building my new business, and, in actuality, I was often distracted from my family because I was so focused on building this business. I stuck my life savings into this business, and I wanted to make it work. However, now, I was going to dialysis 3 days a week, which for me at first was a real inconvenience. Dialysis sessions were long, tedious, and could be pretty boring, with not much to do but hang out while the dialysis was being completed. I started thinking that I should use these dialysis sessions to my benefit and to help everything go as well as possible with dialysis. Many complications can happen with dialysis. I started to practice learning how to meditate and how to pray during these dialysis days. As a result, I started feeling closer to God, and I grew to have more love and trust. My time with dialysis actually became a time of deep inward healing of my mind, which I firmly believe helped me physically during the 8 months I was on dialysis.

Initially, I was going to try to work during the dialysis sessions but couldn't do so because it just took too much out of me. However, when I focused on healing, meditating, and praying, I was finished with dialysis in 8 months, by finding out my son was a complete match for a kidney transplant.

I am very blessed with four wonderful children. My three older children were matched for a kidney transplant, which is a miracle in itself. My daughter, who was born in 1998, the year this kidney disease was discovered, was a match, but she could not donate her kidney because it was found that she had a kidney stone in one of her kidneys. My second daughter was also a kidney match, but the arteries of her kidneys were reversed, which would have made a transplant very difficult on her, and we didn't want that to happen.

Going through this time was an emotional rollercoaster with who could be a kidney match. A potential kidney donor has to go through a plethora of tests, including psychological tests, to be sure they can handle being a donor. It took months to have the testing done to see if my children were a match. For whatever reason, the testing for a kidney match took months between my daughters, Molly, and Georgia. The process of matching for a kidney for me took an emotional toll on all of us, and I recognize that it took especially a toll on my children. When my daughter, Molly, was found to be a match, we were all so happy and thought we, indeed, would wrap everything up and get started with the kidney transplant. Then, the news of the kidney stone in one of her kidneys dashed our hopes. Once again, we were all on a high finding out the incredible news that Georgia was also a match until we found out about the arrangement of her kidneys in her body.

This time period was so dramatic for me, all my family, and close friends. As well this extended my time on dialysis and delayed the opportunity for the kidney transplant. My steadiness and steadfastness in meditation and praying during this time were vital for me to have resilience, along with trust that all would occur in the right way at the right time.

A couple of years before the kidneys started failing more quickly, I transferred my case to the University of Maryland Medical Center. They also had a comprehensive medical care approach to the type of kidney disease my kidneys suffered from. Going through the process of having a kidney transplant when it's time is very arduous in usual times. During the 2020 pandemic, this became exponentially complicated, stressful, and drawn out due to heavy loads on the hospital to take care of patients with COVID-19 on top of other patients requiring various forms of medical treatment. Everything was difficult: getting appointments set up, relaying information about my medical situation to medical staff, having phone calls returned from doctors, getting required medical tests completed, making sure that all relevant parties involved in my medical care for the kidney transplant were communicating effectively, and dealing with insurance. Without my amazing wife, Loren continuing to diligently help me with all this coordination, getting everything done for the kidney transplant would have been nearly impossible.

Admittedly, there were times when I experienced heightened tension and anxiety during the months of dialysis while testing was being done with my immediate family for a kidney match. There was so much uncertainty. Thoughts would run through my mind: Will my health remain stable? Will we be able to find a kidney match, and if so, how long will it take? How long will I have to be on dialysis? I am often so tired. Will I be able to continue to keep working with my business? Will everything be ok? Many other questions would arise also. During these times, I would remind myself to cling to the hope in this world. I would rely on my meditations, my praying, and my chosen spiritual path for courage, strength, and patience, as well as to help my mind be calm and clear.

As I said, it is so very uncommon to have multiple family members, in this case, my children be a match for a kidney transplant. Yet, the greatest miracle was to be revealed. My son was tested for a match that originally came back, showing he was not a match and could not be a donor. However, after months, a doctor noticed that his DNA was the same as mine, and the bloodwork showed that our blood type was the same. He ordered another test, and it came back that we were a perfect match. I feel like this was a blessing, and through this blessing, my 21-year-old son, Jack, had the courage to donate his kidney to me.

The day in May 2021 that we both were getting ready to have kidney transplant surgery, me as the recipient and him as the donor, was one of the most humbling experiences of my life. Watching my son being wheeled out of the room to have to go through a serious procedure to save my life was very, very hard for me. I lay in the hospital bed feeling helpless.

Since it was COVID, only Loren and my daughter, Georgia, were able to be in the hospital waiting room during the surgery. Usually, there was only 1 person allowed for a patient receiving surgery. Technically, there were two surgeries being done: Jack's surgery to harvest one of his kidneys with his surgical team and my surgery to transplant his kidney to me with my surgical team. Loren was the person for my surgery, and Georgia was the person who had Jack's surgery. Georgia and Jack are fraternal twins and are both very connected and close to each other. Molly was home with our youngest daughter, Lucy. When Jack got into recovery, Loren texted Lucy and Molly, along with my mom, my brother, my sister, and my in-laws, to say that everything went well. Loren and Georgia would be able to see Jack in a couple of hours once he got into the hospital room. Everyone had been waiting for this news after eight long months of dialysis, and the day had finally occurred; the kidney transplant was completed and successful! Everyone was overjoyed.

Jack had been in surgery for about 2.5 hours, donating his kidney to me as the surgical team prepared to harvest it for me. I was in surgery longer as the medical team attached Jack's perfect kidney to one of my kidneys. In a kidney transplant, the recipient's kidneys are not removed. Instead, the newly harvested kidney is attached to one of the existing kidneys. The newly transplanted kidney is then "hooked up" to work with the body. When I was in recovery, the lead surgeon shared with me that the new kidney worked immediately as soon as the team started the process for the new kidney to function. It is truly an amazing surgery.

When I woke up in the recovery room and became more fully aware, I said to the recovery medical team, "Man, I have to pee."

They replied, "Go ahead."

Again, I repeated, "I have to really pee."

They again said, "Go ahead."

In my mind, I thought that the surgical team had not started surgery yet. Slowly, I came to realize that the surgery was done and completed. The recovery medical team said to me that it was a really good sign that I had to pee. Once I became more awake and aware, I could feel the kidney work, and it was a remarkable feeling.

Loren still had not heard any news about me for several hours. Somehow, there was a communication gap, and no one had shared with her that I was already in recovery. When I was waking up from the surgery, the lead surgeon came into the room to check on me with a smile on his face. A thought came to me, "Does Loren know I am in recovery?" I asked the doctor, "Does my wife know that I am in recovery?"

He replied, "I don't know, but let's call her and let her know."

He took out his phone and called her phone number. He relayed to Loren that I was doing well. He also reiterated that Jack was in recovery as well. The surgeon put the phone to my ear, and I told Loren that I was doing okay. I also asked her about Jack. Unknowing to me, Jack was in recovery, not very far from me, actually in a room across from me.

Due to the restrictions in the hospital from COVID-19, Loren and Georgia were not able to see me or Jack while we were in the recovery rooms. In Loren's words, it was "heartbreaking" because "her two guys" were undergoing such a pivotal surgery together, and all that she and Georgia wanted to do was to "See us and put their eyes on both of us." They could only see us when we arrived in our hospital room after being in recovery. Jack got into our hospital room first, and Georgia went with him. Once I arrived in our shared hospital room, Loren was able to come join us. It was certainly a very special and wonderful reunion after such a long, intense day. I was so proud of my son, who had shared such a life-giving gift with me.

The great news was that my son, Jack, came out of the surgery wonderfully. He recovered very, very well and very quickly. For me, it was the same experience. The very next day, Jack recovered so quickly that he was discharged at around 2:00 pm to our house. Georgia got him home and comfortable and was taking care of him. I was able to get up and walk around for a little while. The hospital room was, in

Loren's words, "Grand Central Station," with all the monitoring going on of my body's systems and the education the medical staff were giving.

Just two days after having kidney transplant surgery, I was discharged home. I was so excited to go home and be with my family. Immense feelings of joy and gratitude consumed me. Of all the patients who had received a kidney transplant at the University of Maryland Medical Center, I was the patient who was discharged the fastest ever. It was a record-speed discharge from the surgery center. Once we were home, both Jack and I focused on our complete healing, restoration, and then resuming our lives. A huge highlight of getting home that first day, besides being with my family, was the shower. During the time that I was on dialysis, I was unable to take a shower due to the port of the dialysis. This meant no shower for eight months! I could only sponge-bathe to wash my face, my hair, and my body. As soon as I got home, I got into the shower, and it was bliss. My body felt so incredibly refreshed and clean. To this day, I never take a shower for granted.

This entire journey with the kidney disease was my experience flirting with the saying and belief that "Ignorance is bliss," which is truly a myth. This journey with "Ignorance is bliss" was going to now tie into my future. What we didn't really know or didn't realize, or nobody saw coming, was the storm that was on the horizon. The end of this road with the kidney disease that culminated with a kidney transplant was just the preparation.

Chapter 2: Ignorance is Bliss Strikes Again

My kidney transplant was completed on May 10th, 2021. I thought this would for sure be my only major health challenge. But nothing could be further from the truth. Fast forward to March of 2023. It was early in the morning on March 14th, and I was taking a shower to prepare to go to work. Without warning, I fell uncontrollably in the shower. My wife, who was in bed, heard me fall and came running into the bathroom to check on me. She asked in a terrified voice, "What happened?"

My calm response to her was, "I slipped."

However, she did not buy my reasoning. She said, "I think you had another stroke."

After 27 years of marriage, I had enough sense while lying flat on my back in the shower not to argue with her. She called 911 immediately for the EMTs to come check me out. Meanwhile, in my mind, I was thinking to myself, "I have to get dressed to go do a training session."

I really didn't think I had a stroke because I wasn't dizzy, and my speech wasn't slurred. Unfortunately, when the EMTs came and checked me out, they confirmed that I had a stroke and took me to the hospital by ambulance.

This time was quite different from the previous time I went to the hospital. Once again, I was checked into the emergency room, and a CT scan was conducted that confirmed my worst fear – I had a stroke. Loren had called my mom saying that I had had a stroke. My mom, who lives in Florida, breathlessly and emotionally contacted both my brother and sister to inform them. My brother, Michael, was also living in Maryland, and as soon as he found out, he texted Loren and tried to reach her. He wished to come and see me immediately at the hospital. He started driving towards the hospital from where he was and was able to connect with Loren through text on the way. When Michael arrived, I was pretty quiet as I was trying to process my situation and was saying prayers under my breath.

Life in this country is quite busy, and we have not seen each other for some time, even though we live about 30 minutes from each other. Michael came in and said, "Hello, Matt."

Loren had told him that I was not very aware of my left side. He came to sit next to me on the left side of my bed and held my left hand. Michael's voice was calm and very reassuring. He was adamant that I was going to be ok and that I was going to recover fully. I turned my head towards him, and he said he had no doubt I was going to get better as he kept holding my left hand. Then we were just quiet for some time, with him sitting next to me and me lying in a hospital bed.

I deeply needed to hear what he was saying. His words and presence were very important and healing for me.

This hospital event turned out to be now the beginning of traveling on a new unchartered road with a new health challenge that started over 3 years ago with dialysis, followed by the kidney transplant to this present day. Presently, there are still residual effects from the events that occurred in March 2023. The first effect after this stroke was that I really couldn't walk, and my left side was not functioning properly. My left leg and my left arm specifically were not moving well. I could not do the most basic daily functions because the left side of my body was simply not doing what it was supposed to be doing.

Yet again, I was focused on how I could get out of the hospital as fast as possible. This time, the doctors and nurses were very worried and cautious about how to handle me. I was admitted to Carroll Hospital, and I found that my spiritual journey had really taken off. I deeply desired to find meaning in why this occurred again. I also had a deep desire to feel spiritually connected to God and to be supported, protected, and guided. I made a point to pray every day and to thank God for the time I still had on this earth. I faced the threat of departing this world from the possibility of kidney failure. So, after having yet another stroke, I was incredibly grateful to still be here, living.

I was admitted to Carroll Hospital for a couple of days. The first night, there was a tense exchange with a nurse over a urinal. I wanted to go to the bathroom, and the nurse told me that I needed to use the urinal instead, which I could not do without making a mess. I wanted to get up and walk to the bathroom, but this was not allowed. So, we bantered back and forth about what a urinal is. I said that a urinal was attached to a wall that I could just walk up to and use. She was holding the plastic bottle, saying, "This is a urinal."

I said, "This is a bottle. Why don't you just get a Pepsi bottle and call it a urinal? What's the difference?"

The nurse then gave me the urinal and left the room. There was a nursing tech still in the room. I turned to the tech and said, "I am not using this," and gave the urinal to the tech. At the time, I just could not really understand the seriousness of my medical situation. Using a urinal made me feel degraded, and I was angry.

The next morning, the physical therapist came to evaluate me for mobility and walked with me down the hallway. I was walking quite well. Even the physical therapist (PT) said I was walking well. I desperately wanted to brush my teeth as they had not been brushed for over 24 hours. The nursing staff gave me a small toothbrush and toothpaste. When the PT showed up, I shared that I wanted to brush my teeth. We were walking down the hallway towards a water fountain. We got there, and I started thinking, "How am I going to brush my teeth due to the paralysis and weakness on my left side? How do I do this?"

I thought that I had to shift my weight to my left to stand at the water fountain as I was preparing to brush my teeth. Suddenly, I lost complete balance and fell very hard on the floor. The physical therapist had walked away for something because she thought I was doing so well. Therefore, no one was right beside me to catch me as I quickly lost my balance. As an added insult, I fully peed myself after hitting the floor so hard.

Nurses that were down the hallway saw me falling onto the floor and came running and rushing toward me. I fell due to weakness and paralysis on my left side, which included my left arm and left leg. Later, it was found that I also had a visual perceptual deficit called Left Neglect, which also definitely contributed to the fall. Left Neglect occurs when there is damage to the right side of the brain in the right cerebral hemisphere after a stroke. Vision is fully intact; however, there is a lack of perception and awareness of the left side of the body, almost like it's not there. The nurses were assessing me and said to me, "You must have had a seizure."

I said, "No, I did not have a seizure. I lost my balance trying to brush my teeth." I never lost consciousness. The PT had now returned to the scene, horrified that I had fallen onto the floor. She exclaimed, "He was doing so well!" They helped me get back up and back to the bed. Now, the nurse that I had been arguing with the night before came over to the nurses helping me and said, "See, I told you." All of this was an entirely defeating experience for me.

During my admission, this hospital and my wife, Loren, had been reaching out to the University of Maryland Medical Center to inform them of my medical condition. Both my family and I definitely wanted to be transferred there because this was where I had my kidney transplant. Obviously, having a kidney transplant and now a 2nd stroke was a medically complicated situation. I was transferred to the University of Maryland Medical Center at three o'clock in the morning, basically naked by ambulance.

I arrived at the medical center waiting in a room, still basically naked, while two nurses did a full body check, checking all over and everywhere to be sure that I did not have anything like a gunshot wound or drugs or anything else somehow on me. They kept telling me this was necessary. Here I am, just having a stroke and only wanting the medical attention that I needed, and this is going on. I kept thinking, "This can't be happening." It was a totally humiliating experience. However, throughout the course of having now two strokes, I learned this was part of the process and had to just suck it up.

It was cold outside, and it was cold in this waiting room where I was waiting for a couple of hours with a lot of people walking around while TVs were turned on and blaring. Fortunately, I was covered by a number of blankets and was warm. I was in shock with everything going on. The entire time I was in this room, I was trying to block out this entire experience. I was feeling traumatized from having fallen in the shower at home, having a stroke, being admitted to Carroll Hospital, and now everything happening so far at the University of Maryland Medical Center. Truth be told, I felt very helpless because I could not take care of myself. Words cannot adequately express my dismay, disappointment, and discouragement that I was in this predicament.

Eventually, I was admitted unexpectedly to a private hospital room. My wife Loren, along with my oldest daughter, Molly, who was living in Dallas and flew to Maryland, also now a nurse, came to the hospital to take care of me along with the nurses at the University of Maryland Medical Center. I really must acknowledge the wonderful care that the nurses gave me while I was in the hospital, especially with everything they had been through because of the pandemic.

Now that the pandemic is over, I am back in the hospital. I really could not believe this was happening. I just wanted this to be a dream that I would wake up from. Loren, who is a doula for newborn babies, had stopped working to take care of me. Then there was my lovely daughter flying home from Dallas, taking the time to help take care of me as well and be with me in the hospital. I was truly ashamed of the condition that I was in. I felt that no daughter should see her father in this condition. I also did not want my other 2 daughters and son to come to the hospital and see me for the same reason. Georgia was living in her own apartment locally, and Lucy was at home in 11th grade.

Molly had joyously become engaged to her future husband, Brad, in 2022. Everyone was upbeat and ecstatic for this heartwarming impeding occasion that would be in August 2023. Save the Date invitations had already been mailed. With the arrival of this sudden, unanticipated difficult, and serious health upheaval, the wedding plans were now on hold. A note was put onto the wedding planning website that due to "Unforeseen circumstances, the wedding would be postponed with more details to follow." It was truly sobering that Molly's grand wedding that she was envisioning would not be coming true. There was uncertainty about my health, and it was decided that the wedding instead would be a small, intimate affair sometime in the near future before I would have to undergo further treatment for the mass that was in my brain. More about that in a moment. Brad, Molly's fiancé, who was also a nurse, generously came from Dallas and stayed for some time to help me in the hospital.

To me, all of this was a desperate set of circumstances that were completely out of my control. I needed encouragement. My daughter would write a psalm out of the Bible on the whiteboard in my room every morning. This helped me cope with the situation that was going on. When any person faces a health crisis, whatever spiritual meaning they have in their life can be monumentally helpful in their healing and recovery. I have found this to be true for myself. I was praying for a miracle to recover from this 2nd stroke very quickly.

My entire family, along with friends and clients who knew what was going on, rallied me by sending texts or phone calls to Loren. I was not using my cell phone at this time. Loren would share with me what people texted to her or relayed by phone when speaking with her. All of this meant the world to me. However, with so much going on and happening so quickly, doubts and various fears would creep into my mind. There were times when I felt very lost. I felt at times that I was losing the person that I knew I had been. On March 17th, my brother, Michael, sent me a text through our family group and asked Loren to read it to me. Perhaps he knew that with so much unpredictability with what was happening, I felt deeply insecure at times. His text was this:

"Good morning everyone. This message is for Matt. I know Loren will relay this to him. Matt, you are in our hearts, minds, and prayers. We love you. We are looking forward to you coming home. You're an amazing brother, son, husband, and father. You're a selfless, loving, dedicated, successful man to whom people look up. Many people are praying for you. You will recover. Your strength is amazing. You're always on our minds. Much love from all of us."

Loren shared this text with me, and I was reminded again of who I was, what I had been, and what I would continue to be. Again, I reaffirmed to myself that I would be the best person that I could possibly be. There was a wish that, hopefully, in the future, I could help other people become the best they could be as well.

I had a whole team of doctors that were checking on me daily. However, I was really frustrated just lying in bed all day. I was ready to get up and start moving to begin beating this stroke that had happened. Unbeknownst to me at that moment, this process of healing from now more than 1 stroke was going to be a longer process. I came to realize that nothing happens as quickly as we like it to. I could not find answers in my logical mind how to deal with everything happening. The feeling of helplessness was overwhelming. I felt that I had to continue to deepen my spiritual path. The answer that came was to let go of my logical mind since everything happening just didn't make logical sense to me. I decided to put my faith and trust in God to guide me through this process.

The highlight of my day during my hospital stay was when the nurse would come back with my labs from my bloodwork to share with me how great my kidney function was. The truth is, if I did not go through the kidney transplant, I don't think I would have been able to handle everything I was going through at the hospital. Then, the dreaded day came when a CT scan and MRI showed a "mass" in my brain. This was also seen at Carroll Hospital.

The first MRI scan at the University of Maryland Medical Center was horrible. It was done at night. I had to have a contrast injected into my right arm so that a lot of detail could be seen in the MRI, including blood vessels. Usually, the contrast injection is done before going to the MRI lab. For some reason, this did not happen, and instead, while in the MRI machine, the MRI technician tried to repeatedly inject my right arm with the contrast. This was incredibly painful. The technician kept telling me to hold still. I responded, "I am trying to hold still! But I can't hold still! Someone is stabbing me in my right arm with a needle!"

I was in so much pain and so upset that if I could have freed my right arm, I would have hit the technician in the face. At a later time during my hospital stay, I saw the MRI technician from that awful night. He admitted to me that the MRI scan process that night was definitely not good. Unfortunately, the MRI scan confirmed the scans from Carroll Hospital.

All my family, including my mom, brother, and sister, as well as Loren's parents, were sad, frightened, and worried about the scan result. How could this be? When the medical team told us this, this is when I decided to step out of the 'Ignorance is bliss" thinking. The medical team felt sure that this was a brain cancer tumor. I am recovering from a stroke and am now getting this news. There was a lot of debate within the medical team: do we give more time for the brain to recover from the stroke, or do we do a brain biopsy to understand what this is in the brain? It was a risk doing a brain biopsy so close to just having a stroke, and complications could happen. My brain was already inflamed from the stroke.

My wife and daughter were being briefed daily by the doctors along with me. The neurologists had shared that all the symptoms I was experiencing from the stroke could recover within six to nine months. Some would recover in days, others in weeks to months. On the other side, the medical team was looking at the MRI. Discussions about a brain cancer tumor called a glioma or lymphoma, doing a brain biopsy, possible surgery, and other scary-sounding medical interventions like radiation or chemo occurred throughout these days by the various medical specialists that were looking at my medical situation. Plans kept changing or being updated with the newest thinking from the consults of the medical team. Thankfully, Molly, being a nurse, was able to understand all the medical jargon and terms that were being said to us. She would explain things to me and Loren. All of it was so confusing to listen to and understand especially since I was not able to think clearly from the stroke. Of course, emotionally, I was astonished at this new potentially dangerous threat to my health and recovery.

The doctors decided for me to have a brain biopsy. My family wanted to know what this "mass" on the CT and MRI actually was. For the sake of myself, my wife, and my family, I knew that I needed to accept that this biopsy was happening. I had to trust that this was the best decision. I also had to trust that the biopsy would be done on the right day and timing for the best recovery outcome. Originally, it was planned for March 23rd; however, the biopsy procedure was bumped. This frustrated Loren and Molly as they wished for this to proceed to at least get an answer as to what the brain mass was. The next day, the biopsy procedure was confirmed.

When the nurses were preparing me for the brain biopsy, an incredible peace and comfort came over me that also completely permeated my body and mind. I felt like my body, mind, and soul were completely calm, and I experienced no fear. I feel like this was a divine experience of a connection to God. In my mind, there is no other explanation for how such a feeling of peace came within me and over me that I never felt before and haven't again since.

When I awoke from the procedure, I still had this incredible sensation of peace within me and around me as I heard the nurses and various sounds of medical machines around me while I was in the recovery room. I truly felt everything was going to be OK, which was exceedingly very good because, as you'll be reading, the challenges didn't stop. This abiding feeling of peace and calmness continued to stay with me. When my wife Loren came into my hospital room to see me after I woke up from the biopsy after being in the recovery room, she marveled and made a comment that I was magnificently glowing and could see that I was so peaceful. At this time, I said to myself, "I'm going to speak about this and share this experience." I decided that I wanted people to know that when you have an abundance of some sort of deep spiritual connection, it helps to open a feeling of faith and confidence within you that anything is possible.

For example, here I am, someone who has dyslexia, writing this book who never before thought this was possible. We, as human beings, often live in the limitations of our own mind creation. With each health challenge that has presented in my life, my mind's limitations have been pushed against and confronted. Is what I have been cherishing as truth really truth or not? Is there any other way? Does it have to be this way, I am being told or can it be different? I have found that as I have deepened my spiritual awareness through these health situations, my mind has expanded into possibilities that were not seen before or thought possible.

Up until the time of the biopsy, for me and my family, my hospital stay was filled with uncertainty, apprehension, and uneasiness, especially around what was sitting in my brain. Everyone was so relieved that the biopsy procedure had gone well and that I was recovering without any complications. Then came the best result and unexpected possible outcome from the brain biopsy. The medical team was wrong. There was no glioma or lymphoma. Instead, this cell mass was called a germinoma, which is a very rare brain cancer found in adults.

This usually occurs in children. The prognosis has a 90% survival rate and non-occurrence, and no more surgery was needed. Only radiation would be required, and there was not any urgency to do anything at this time. I would have time to restore, go to physical rehabilitation, and heal from the stroke. This brain biopsy result was a stunning surprise. Even though there was the deepest wish that this would not be any type of cancer, this was truly the best news possible in this situation. Radiation seemed like it would not be a big deal. People I talked to said it would be easy. I would find out later while undergoing radiation that "no big deal" and "easy" would not accurately describe what occurred.

Chapter 3: The Difficult Journey to Return Home

There was such tremendous relief that the biopsy was over when I started to become aware in the recovery room, along with the continuing peace and calm that was throughout my body. As I was in the recovery room, I heard a woman next to me crying. She had just had a tumor removed from her breast, and she was so distraught that she felt flat-chested. She was talking to her sister on the phone. The old Matt would have sat up and said to her, "Are you crazy? Are you worried about being flat-chested? You had a tumor removed from your breast. Why are you worried about being flat-chested?" The Matt who was now emerging within me instead felt empathy and compassion for her as well as for her sorrow and suffering. I lay in the hospital bed, mentally praying for her and everyone else who was having to face difficult surgeries.

Then, I had a repeat vision that I had earlier in the hospital stay here. In this vision that appeared in my mind, I was standing in front of thousands of people. I was talking to them about seeing the great possibilities in their life. I was teaching them that with the right attitude and set of beneficial habits, they could accomplish possibilities they never thought possible. And that when the right attitude and habits were aligned, nothing could actually get in their way. They would achieve success with whatever possibility was manifesting. In the vision, I was also sharing that many well-intentioned people in our lives often throw roadblocks into our path that hinder our power, ability, and foresight to actualize an opportunity previously unseen or unknown to us.

Once back in my room, I was in good spirits and deeply thankful to hear the result of the biopsy pathology that Loren shared with me. I continued to receive occupational therapy (OT) and physical therapy (PT) throughout the hospital stay. Loren and Molly would bring good food from home for me to eat and remain by my side. They showered me with love. Throughout this entire ordeal, since day one started at Carroll Hospital, I felt so much extraordinary love around me. I knew that this was from all of my family as well as friends, clients, and many others whom I did not know very well personally or at all who had me in their minds. As my family and friends shared my story of what was happening to people they knew, these people were much moved and would send their own positive thoughts and/or prayers. All these wonderful people were wishing and praying for me and my full recovery. To this day, I still have a memory of tremendous love that made me feel like I was riding on a pair of wings, propelling me onwards as my travels on this path progressed.

I was cleared to leave the University of Maryland Medical Center and progress onward to a rehabilitation hospital. I will be discharged to Keenan Rehabilitation Hospital in Baltimore tomorrow. Kernan Rehabilitation Hospital is also known as the University of Maryland Rehabilitation and Orthopedic Institute (UMROI). Upon hearing this news, I began collecting the items that were in my room to prepare for leaving tomorrow.

Before leaving, however, I also had to have a PET scan to make sure there were not any abnormal cells floating anywhere else within my body. This time, the contrast was injected into my right arm in my hospital room. The technician was cordial, very kind and very nice and explained everything. This was a very pleasant change from what I was used to when getting scans done. The transporter came to get me in my room and wheeled me down to the familiar basement, where I had the other MRI scans completed. With all the tests that had been completed thus far, all I had heard was bad news. I just kept thinking that everything would be ok. I kept myself as calm as possible. The PET scan took about 10 minutes, and shortly after being completed, the results showed that everything was clear. What an immense relief! I was brought back to my room and went to sleep.

The following day, on March 28th, when I woke up, the EMTs were there to transport me by ambulance, which seemed to be my only mode of transportation at this time, to the rehabilitation facility. This day, March 28th, was a very special day. It was the birthday of my father. Several years earlier, after surpassing his birthday by a few days, he left his body and departed this world on April 2nd, 2016. During the last 2 ½ years of his life in this world, he was diagnosed with Multiple Myeloma, a type of blood cancer. Being discharged from the University of Maryland Medical Center to rehabilitation felt on this day like somehow there was some sort of blessing being bestowed on me. Ironically, our second dog, Scout, a Westie, that we got for our youngest daughter, Lucy, was also born on March 28th and was six years old on this day. We were leaving this hospital soon, and I was so excited. Loren was relieved too and sent a text out to the family at 1:00 in the afternoon saying, *"Just got the official news that Matt will move to the rehab center at 3:00 today. Happy Birthday, Mike!!"* Everyone was thrilled and delighted to receive this news.

When I arrived at the rehabilitation hospital by ambulance, I wanted to get off the transport cot and walk in. The EMTs insisted that I stay on the transport cot. Similar to the University of Maryland Medical Center, when I was wheeled into the waiting room to be checked in, I became aware that I was going to be repeating the same unpleasant experience that happened when I was first admitted there. I closed my eyes and dealt with the full body check. This time, they were much faster, and it was easier. Even with all that was going on, I was happy to still be on this earth.

I thought for sure that I would have to share a room with another person staying in the rehabilitation hospital. Instead, I had my own private room, which was such a huge blessing. Once I got into my room, a nurse showed up, wheeling a video camera into the room. It was placed in front of me above my bed, and every movement in my room was recorded 24 hours a day. I called it "The Eyes in the Sky." I just accepted that "The Eyes in the Sky" would always be watching me.

Shortly after the installation of "Eyes in the Sky," I was confronted with the urinal situation once again. This time, I did not fight having to use the urinal because if I refused, they would have to catheterize me. Of course, I did not want this to happen, so I accepted the urinal. I was successfully able to use the urinal. I learned that I had to check my pride at the door whenever I was admitted to a hospital so that the medical professionals could do what they had to in order to take care of me.

During my official first day in rehab, Loren sent an update by text to our family, *"The rehab center is wonderful. So much nicer. They put Matt through the paces today, assessing him. The therapy plan begins tomorrow. He loves all three of his therapists. Molly flies out tonight."* We had done it! We had made it through all the medical uncertainties at the University of Maryland Medical Center and now victoriously arrived at rehab! I was ever so grateful to Molly for the love and assistance she had provided. Now, I was in a new phase of the trajectory, restoring my health, and it was time for her to go back to Dallas and resume her life.

I was admitted to the stroke recovery unit. My body was medically stable, and most remarkably, my blood pressure after the stroke was great, and there was no need for medication. Ever since the kidney disease was discovered, there was always high blood pressure, and medication was required. This was an amazing change that had no explanation. I was very excited to be at this rehabilitation hospital as I was ready to get up and get going so that I could finally get home. This feeling stayed until I saw the daily schedule for my therapies. I would have daily sessions of Physical Therapy (PT), Occupational Therapy (OT) and Speech-Language Therapy (SLT). It was then that I realized that this was going to be hard work and not a vacation. I was in therapy for roughly 3 hours a day. My daily schedule was:

8:00 am OT for one hour

9:00 am Speech Therapy for one hour

10 am PT for one hour

OT consisted of getting dressed and ready for the morning, along with coordination and games with my left hand. I had to do a 9-peg test with my left hand quickly. We also worked on difficult eye coordination exercises. The different games that I did with OT were not easy due to my condition, and there were days that I was very depressed.

In speech therapy, I engaged with memory and cognitive exercises. I found that my memory was not as sharp as it had been, and this was troublesome for me. Prior to all of this happening, my memory was very strong. Remembering things word by word was one of my strengths. As part of my business, I actively listened to build trust and relationships with my clients. I would listen to my clients and then repeat what I heard them say back to them to be sure that I understood what they were conveying. "You are experiencing this that I am hearing. Do I have this right?" Now, my thoughts were disorganized, and my memory was impaired.

I began to come to the realization that I really was not in control of anything, even though we are under the illusion that we can control how things go in our lives. I started to settle down, and this made things more peaceful and easier for me. Especially at the beginning of my stay here, nothing came easily for me, and it was very depressing. As therapy continued, I realized the therapists were gradually increasing the difficulty of the therapy tasks that I was doing with them. They were doing this so that I could accomplish the easier things first and then continue forward to the more difficult ones. This strategy of progressing the therapy and challenging me helped me to gain more confidence. I was able to see the progress that I was making more easily. As I saw myself doing well in therapy, I also built confidence that I was assuredly getting better and recovering my physical and mental abilities.

Physical therapy was my favorite. We, as humans, like what we are good at. Earlier in my life, I had been a personal trainer, spent many years working out in the gym, and did other types of exercises regularly. While in PT, I did a lot of walking, pulling a sled behind me that was attached to a band around my waist. I would carry a tube of water in my arms while walking around. I learned to walk again. I walked upstairs and walked on a treadmill. I repeated out loud daily while walking what my PT had taught me how to walk with my left side: "Hip, heel, foot. Hip, heel, foot. Hip, heel, foot."

Even after 3 hours a day in therapy, after a rest in my room, I would want to go back to the gym to do some exercises. There was a special nurse who would come to my room and let me walk to the gym, where I could do more exercises. She did this daily, even if I was not her patient that day. We would leave the wheelchair in my room, and I would walk to the gym with this nurse next to me. I was very grateful that I had the opportunity daily to walk with her to the gym. Her kindness and dedication to assist me get stronger and better inspired me.

As my walking continued to progress, I was cleared by therapy to walk back and forth and go to the gym on my own. One of the nurses who was taking care of me in the morning knew that I was walking back and forth to the gym on my own quite often. One day, as she was getting breakfast out for me, we started a conversation. I mentioned to her that during my frequent trips to the gym, I saw many people there who were worse off than me. I mentioned to her that I was self-conscious about zipping around the gym as I was doing in comparison to everyone else.

This was a no-nonsense nurse; she encouraged me and said it was good that I was doing this. She said that I was not showing off, and instead, I was helping others. Being an old-school nurse, she said, "Hey listen, no pain, no gain." I really did not believe this saying because progress, whether in therapy or in our life, can be "gained" free from "pain." However, I knew she was trying to help and motivate me, and I appreciated it. After talking to her, I realized that I could help inspire other patients to do the best that they could, like in writing this book, I could inspire others so that they can enjoy all that this world has to offer them and not limit themselves. I was one of the youngest people in this rehabilitation hospital at this time, and this impacted me. I wanted to do the best possible and worked to do this daily.

One day, while in the gym doing my usual afternoon "workout," I noticed a guy in a wheelchair getting pushed up to the indoor track. I was thinking, "I wonder what he is going to be able to do at the track. By the way, he looks, he may not be able to do much." To my surprise, he lifted himself out of his wheelchair and then proceeded to walk. To see him walking as he was doing invigorated me. Watching him was a potent reminder that many people around me in this rehabilitation hospital were much worse off than myself. Each time I began to feel sorry for myself, I remembered this and was able to shift my mind back to focusing on doing the best I possibly could.

So many people generally think that everything happening now, whatever it may be in their life, is the way things are going be. There is little contemplation or testing of the current reality if this is actually true or not true. Even if things are difficult or frustrating, having a feeling that there is abundance all around us is very important and invaluable. Abundance allows an openness of possibilities. Feelings of hopelessness, helplessness, and isolation do not germinate or grow when we have a perception of abundance. We should carry abundance and its possibilities everywhere we go, no matter what the situation or conditions are. It's also incredibly crucial to have people around you that will support you and not limit you through their insecurities and limitations.

Loren and Georgia would come together each day, see me, and watch me in PT. They would come in the afternoon and stay until the evening. Loren had started working again and needed to take care of all the appointments, paperwork, and insurance. It was mind-boggling how much had to be followed up on, and all of this was demanding and burdensome. I was acutely aware of how much she was doing and having to take on while also returning to her company. I worried about her becoming completely exhausted. The life of being a caregiver in such a situation as mine was overwhelming and required great love, strength, determination, and courage. Loren possessed all these qualities and many others that made her an exceptionally wonderful caregiver. This was such a blessing in my life, and I often would call her my angel.

A bright spot at Kernan Rehabilitation Hospital was the volunteers. These were people who had experienced a stroke and were previous patients at this hospital. I deeply appreciated their positive attitude and kindness. I was so impressed with the volunteers that I myself wished to volunteer in the future. I still have this goal and look forward to attaining this goal when I am prepared and able to become a volunteer.

I consistently thought about my business all the time. The core of my business is leadership and sales training for existing businesses. How quickly can I get back to my business? When will I be able to schedule training sessions? When will it be possible to get back in touch with clients? I am an entrepreneur, and when I was working, money came in to support my family and me. Now, being unable to work was a real hazard for me, and my family's financial stability and financial strain was debilitating. I had been in touch with a few clients who had heard what had occurred to me. They were deeply concerned and sometimes became choked up, saying they were surprised that I was still here on this planet. I absolutely wished to restart working with my clients. Truthfully, I was not in any physical, mental, or emotional condition to engage with any semblance of working with clients or my business.

The rehabilitation hospital had very poor cell service, so it was not possible for me to use my cell phone. This made me quite isolated from the outside world beyond the hospital walls and buildings. In one way, this was helpful because it assisted me to be more focused on my purpose for being in a rehabilitation hospital. I also had a feeling of more peace to just deal with what was going on. However, it was only possible to use the phone in my hospital room. People could only reach me by this phone if they had the rehabilitation hospital phone number and if I was in the room. This was a downside of not being able to be more connected to those people close to me. I very much wished to speak with my brother, Michael. We had not been able to talk since my admission to Carroll Hospital. I asked Loren to reach out to him and give him my hospital room phone number. Michael and I were able to reconnect and talk several times. His words and positive energy boosted my confidence and motivation.

My mother in Florida had not been able to be in direct contact with me since the Carroll Hospital admission. Loren was sending her regular updates and talking to her about what was going on with me and how I was doing. Mom was trying to be positive. However, she remained extremely concerned about me. She was unable to sleep much at night and persistently thought about me. She was profoundly sad about me and my situation. On April 3rd, my sister received a phone call from my mom. "I am in the ER. I fell backward onto the tile floor, flat onto my back, and hit the back of my head very hard. There is a very large bump on the back of my head." She was being evaluated in the emergency room to see if scans or X-rays may need to be done. Upon hearing this shocking news, my sister was urgently seeking to find out the extent of her injuries to see if she had to book an immediate flight to be with her.

Mom was going to a funeral at her church, where she also volunteers in the office. She was dropping something off in the office at the counter and turned around to leave. Somehow, as she was turning, her shoe got caught on the floor, resulting in her losing her balance and falling backward directly onto her back while also smacking the back of her head. The shocked office staff immediately rushed over and called 911 for an ambulance that transported her to a local hospital emergency room.

Astonishingly, she had no skull, neck, or back fractures, no concussion, and no bleeding in her brain after such a traumatic fall. The bump on the back of the head was very sore. However, it was only deeply bruised and nothing else. In fact, the emergency room doctor said to her, "You are the healthiest person in the ER today." There was no need for X-rays or scans. A friend took her home and stayed with her overnight and the next day to be sure she was all right.

She had bruising and muscle stiffness on her back, her neck, and the back of her head. Nevertheless, she was able to recover quite easily and quickly. My mom and all of us believe she was spiritually protected from sustaining much more serious harm from this fall. The outcome of having such a fall could have been disastrous. After this fall, my mom decided to let go of her worry and trust more deeply that God was blessing and helping me. She became calmer and was able to sleep better and be sure that she was eating well. Michael reached me by calling my hospital room phone about Mom when the fall happened. I was relieved to hear that she was okay. Mom and I were able to connect and talk over the phone. She was happy to hear my voice, and I was happy to hear hers.

My rehabilitation hospital stay was solely focused on healing from the stroke. I was directing as much energy as possible towards my healing and my health. My room was near the nursing station. I could hear doctors and nurses talking about me outside my room. This was the worst part of being here. I could not stand hearing them talking about me. They were talking about the "mass in my brain." Other times, the doctors and nurses were preparing for patient rounds and were discussing my case outside my room. All of this was a painful reminder that this rehabilitation stay and recovery from this stroke were just a part of everything going on. More was to come, and there was more to deal with around this mass in my brain. It was not totally clear what the medical intervention was going to look like, and I was questioning if I was going to live.

I am a private person, and being in any hospital was utterly not private. There were constant questions asked that I had to answer. Being in a hospital and having so many questions thrown at me felt like a violation of my privacy. I knew their intention was not to pry into my life; however, for me, it felt like my life at the moment was an open book to many people, whether or not I wanted it to be.

I also experienced a lack of privacy with how I was dressed at times in the hospitals, especially when interacting with medical professionals. This consistently occurred throughout my hospital stays. For example, towards the end of my rehab stay, 3 nicely dressed doctors, while I was dressed in a hospital gown, would come to see me in the morning. Each time, I would feel very uncomfortable being dressed this way in front of them. I realize that they were not consciously trying to make me feel less dignified in this situation. However, this was occurring, and it was a subtle power play of who was in control and who was not. Me, not in control, and them, in control. I found the lack of privacy and resulting feelings of intrusion to be the hardest and most unpleasant aspect of being in these different hospitals since March 14th. I deeply desired to have more privacy when it was possible.

One of the occasions of these 3 doctors doing daily checks was checking the wound on my head from the biopsy. Each day, they would stare at my head wound and debate if the stitches would dissolve or had to be surgically removed. One of the doctors was a previous surgeon. The nurse practitioner was particularly concerned that the stitches would not dissolve on their own. The doctors eventually came to the conclusion that they would dissolve. I shared this with the nurse practitioner when I saw her, and she stated, "Doctors don't know anything." Fortunately, she was wrong; the stitches dissolved, and the head wound healed.

Later in my rehab stay, I began walking around the rehab facility more often with my wife, Loren, and my daughter, Georgia. They often came during the afternoon, and we would go for walks, including to the dining area and sometimes even outside. I started to think it would be great to be discharged sometime soon. After a few days, the news came that I was going to be discharged. In order for this to happen, we had to have either in-home therapy or outpatient therapy set up for me before going home. To tell you the truth, I was super excited.

Suddenly, the news came that I was going to be discharged the following day. On April 6th, Loren texted the family, *"Matthew is coming home tomorrow!"* Loren was able to work out the details for my outpatient therapy, which was recommended by the rehab team as opposed to in-home therapy. I felt like, "We're now making progress!" Inwardly, I thanked God for protecting me and taking care of me through this process. I was now ready to take on the mass in my brain.

I was looking forward to not having this mass in my head any longer. Loren took care of scheduling medical and therapy appointments that needed to be set up after I got home. It was a lot to contend with getting all the appointments organized and into place. I did feel overwhelmed thinking about the dizzying amount of medical consultations and therapy sessions I would need to attend. However, I was enthusiastic about being home with my family and my two dogs.

When the next day came, the day that I was finally leaving this place, I was extremely happy to say to my therapist, "I am going home and am going to be sleeping in my bed tonight." Loren, with Georgia, came to pick me up in the afternoon, and I was proud. In fact, I was very proud to be able to walk from my room to the outside and to the car with my PT. Loren and Georgia were also by my side, and it was a lengthy walk to get to the car. It was hot outside, and I was able to do this, which was an enormous accomplishment considering where I was when I first arrived at Kernan Rehabilitation Hospital. Prior to leaving the rehab hospital, I had practiced and learned how to put myself into the car. Now, the rehearsal practice to get into the car was over, and I was getting into our car and going home. It was an incredible feeling of relief and joy as we were driving away from here towards home. I was so grateful to God that I had been so protected and now was going home.

Chapter 4: There's No Place Like Home

As I was riding home on the passenger side of Loren's car, I thought about the new life l am now blessed with. I had weathered a severe storm that threatened to damage me greatly. Although I had residual effects from the stroke, there were a lot of functions and abilities that I still had intact or that were improving. For this, I was extremely appreciative and felt blessed. When I got home, we began unpacking my clothes from the past three weeks in the hospital. It felt so good to be home; as they say, "There's no place like home!" It was April 4th, and Loren texted the whole family, *"We are all home. Matt had a nice early evening nap, then a taco dinner now. Getting ready for a shower and then sleeping in his own bed. Hard to believe he's been in the hospital for 24 days. Good to be home."* Everyone was elated with the good news and was so thankful that I was home. I felt beholden for the protection that I received from God during this turbulent time.

Being home, although wonderful as it was, was also a stark reminder that the outside world had not stopped. Bills needed to be paid, and the responsibilities of being a father and a husband hit me hard. I realized, "Wow, how great it feels to be home. Even with many responsibilities, I'm grateful to be home, and I wish to be able to take care of my family." During this time, I had no idea what it would take to recover fully. At that time, I wondered, "How am I going to manage everything—attending therapy appointments and finding time to rebuild my business?" It should have been obvious that I needed to focus on my health first in order to restore my business. Yet, I was still in the mindset that I would bounce back quickly. I've always been able to recover quickly throughout my life, and even after everything that's happened, I believed I would once again return swiftly and get back to my business. What can I say? I'm a slow learner.

I and my family still did not know what the medical course was going to be to remove this mass from my brain. It has been so good to see my family and 2 dogs and be back home with them. Even so, there was the unsettling lurking of this mass in my brain. I found out that I would need to get a medical procedure called a Lumbar Puncture. A Lumbar Puncture is also known as a spinal tap. It is a medical procedure where a needle is inserted into the spinal column to draw out spinal fluid. This spinal fluid circulates around the brain and spinal cord and is called Cerebral Spinal Fluid (CSF). This procedure is done to see if there is any infection or abnormal cells in the fluid. My oncologist told me that I needed this procedure done to determine what medical intervention would be done for this mass in my brain.

This outpatient procedure was supposed to be quick, easy, and not painful. In my reality, the complete opposite was true. I met the oncologist when I arrived, and another specialist doctor entered the procedure room. I was sitting on the edge of an examination table, leaning forward. The specialist, now behind me, inserted the needle into my lumbar spine and was unsuccessful in withdrawing the spinal fluid. Anyone who said to me that this was a painless procedure must have never had this procedure done on themselves. This was absolutely not pain-free. The specialist tried 2 more failed times to extract the spinal fluid through the needle. Each time that he tried to take out the spinal fluid from my lumbar spine, it became more and more painful.

Fortunately, the oncologist had reasonable sense to see that I was in a lot of pain. This obviously was also not working. She said, "Enough. This is going nowhere. We are going to do this in the OR (operating room)." I was startled upon hearing this and was not ready for any operation. She clarified that there would not be an operation. She said that I would be lying on my stomach in the operating room, and they would have precision instruments available to draw out the spinal fluid. The lumbar puncture was scheduled for next week. Needless to say, this experience was difficult, uncomfortable, and physically hurtful. I wasn't looking forward to repeating this again. However, once again, I was reassured that this would not be painful and would now be easy to do in an operating room.

Next week arrived to face this again. Loren was waiting for me in a private recovery area while I was in the OR lying on my stomach. As I lay there on the OR table, unbelievably, this medical team was again unable to draw any spinal fluid. Once again, this is not easy, not quick, and very painful. While I am laying there having this lousy and again painful experience, incredulously, the cell phones of the medical team are going off. While the calls are being answered, I think, "I can't believe they are on cell phones while I am lying here. This is how cell phones are ruining society." As if I am not there or unable to somehow understand what is happening, I hear them talk about me. A conversation all around me, about me, bizarrely without including me. As I have mentioned previously, this was very intolerable for me when this would occur. What's the deal? Why do they seem always to do this? Do they not know that I am hearing every word they say? Is it too much to include me, their patient, and that they are doing all this to, in the conversation?

The current situation's lack of success is leading the medical team to think I would have to return to do this again. In my mind, I thought, "There is NO WAY I am doing this again!" From the start of being in the OR, which seemed to turn into a living nightmare, I had been praying very hard for this to get completed.

While lying on my stomach, I realized that I was in pretty bad shape. The medical team was unsure how to get the spinal fluid sample. I came up with the idea to get into a plank position. As I would hold the position, I would push out the spinal fluid into the needle. The medical team agreed to try this idea. To this day, people ask me, "How in the world were you able to get into a plank position on an OR table, especially after the stroke?" I really have no answer for this except that

I had been praying so rigorously that I somehow had the power to do this. I stayed in the plank position, pushing backwards into my back. I could hear them say, "Ok, we got up to....we are getting the fluid...keep going." Finally, this was accomplished. I laid back down on the OR table.

Shortly after, someone came into the OR and took me into the recovery area where Loren was waiting. I rested awhile before leaving to go home. In retrospect, looking back, I do realize that I had spiritual protection. I came to find out later that a lumbar puncture is not as simple as they say it is and for sure not a complications-free procedure as it was made out to me. The more they inserted the needle into my lumbar spinal column, the greater the risk of complications. It's also hard to explain how I managed to get into a plank position and hold it for as long as I did. I now think there is reason and purpose when medical professionals say, "Oh, don't worry, this is not going to be painful, and it will be easy." If a person fully understood what is involved in many medical procedures or interventions, they would either refuse to do them or be very reluctant.

A few days later, one afternoon, on our way to an outpatient therapy appointment that Loren was driving me to, her phone rang in the car. She quickly put her phone on the car speaker because it was my oncologist, so we both could hear and talk to her. The oncologist had a very peppy voice and shared with both of us that I was not going to need chemo to treat the brain mass! "Mr. Blake, no chemo for you, my friend. There's no further spread, so you'll only need to receive radiation. This should treat the Germinoma in the brain." The cerebral spinal fluid had come back clear. Excitedly, Loren texted the family, *"Incredible News!! They will be treating Matt's tumor with radiation only. It will start right after the wedding, 5 days a week for 15 minutes for 3-6 weeks. Should make a full recovery!!!!! Such incredible news!!!"* I would be getting radiation only, which was very good news at the time.

One thing you learn about being a patient is that many people are very quick to share their experiences or their opinions on what they "know" through their own experience. What I had heard so far was that the radiation wouldn't be a very big deal. What I also found out was that the people who are sharing this with me never went through radiation sessions for a mass in the brain. It was an enormous relief to all of us that chemo was not going to be involved. Once again, my loving God was protecting me because, in the condition I was in, chemotherapy would have been incredibly difficult on my body—possibly even dangerous. Chemotherapy is hard on anyone's body, for that matter.

During my early return back to home, I started outpatient therapy at Kernan Rehabilitation Hospital. It was difficult for me to return to Kernan. I felt like I was going backwards because I was returning to this rehabilitation hospital, even though it was outpatient. I participated in PT, OT, and speech therapy in outpatient sessions 2 times a week. Not being able to drive and having to be reliant on Loren to drive me there was very uneasy for me. Before all this happened, I was living a fully independent life and was able to do all my life activities. Now, suddenly requiring assistance from others, my independent life was drastically altered and was, in fact, missing. I was trying to cope with how my life had shifted so unexpectedly and tremendously, which was challenging. I needed to rest, restore, and sleep as much as I could. My mind and body were overwhelmed by 7 every night, and I had to go to bed. However, sleep was not as easy as it was before all this happened. I often felt like I never slept throughout the night and would wake up feeling tired and unrefreshed. Throughout this time, there were periods of depressed feelings and great doubt.

My daughter, Molly, and her fiancé, Brad, generously offered to change their wedding date to a much earlier date with everything going on. It was also decided that the wedding would be an intimate affair rather than a large gathering. The date chosen for the ceremony was early May, and the bridal shower was to be the day before. Even with all the concern about me, everyone put their energy into making this wedding a truly special experience for the bride, groom, our family, Brad's family, the wedding party, and all the guests. The bridal shower was a delightful gathering for Molly and her special guests attending. Brad, his father, his mother, and his sister visited us at our house before the wedding. We shared a very nice dinner, and I realized how great a family they are together. Each of them is such a wonderful person. I could not ask for a better son-in-law or in-law family for Molly.

Before the day of the wedding, I had been ardently practicing walking so that I could walk Molly down the aisle as perfectly as possible in front of everyone present. I also had been practicing the energy and clarity of my speech as I would be reciting a prayer at the wedding. I was determined to walk Molly down the aisle as father and daughter and deliver the prayer in a confident and clear voice. When the day came, I was so incredibly proud to walk Molly down the aisle. It was a beautiful wedding ceremony that was spiritually motivated. Being a part of their wedding ceremony inspired me to want to be a bigger part of living life and help them and everyone within my family. The bride and groom of a wedding determine the theme of their wedding and how they would like the wedding to be that has personal meaning to them. Molly and Brad decided that they would like to demonstrate their spiritual connection to Jesus, their chosen spiritual path. During the ceremony, both of them accomplished this in their own personally meaningful way. The wedding was upbeat and happy. I found the experience of the wedding to be magnificently amazing. However, I wanted to return to the "old Matt." The "old Matt" independently functions in all ways with a successful business, familiar acquaintances, good friends, and constant social activities. I was unable to do this where I currently was physically and mentally.

A few days after the wedding, I started to get back to work and to my business. When I was hospitalized, my daughter, Georgia, took over my business email account. She handled my correspondence with clients and provided updates regarding my progress and healing. Georgia was instrumental in keeping the sphere of my business clients, associates, and acquaintances in the loop regarding what was happening. She was a necessary thread that helped keep my business "alive" even though I could not do anything actively. Without any previous experience, Georgia had jumped into this vital role for my business. Several clients shared with us that they were impressed with her professionalism and clarity of communication. My clients deeply appreciated the updates she regularly sent to them. I certainly recognized her caring, diligence, willingness, and ability to do this for her father. I felt immense pride in her as my daughter and my "co-pilot" navigating my business during this time. Now, I was picking up where Georgia had left off.

I was reaching out to clients, even seeing if I still had clients after being out of contact for such a long time. I did have some clients still waiting for me to start training sessions again. Before this occurred, I had my own office with a training center through a franchise. In my current situation, I had to make difficult decisions about my business that weighed heavily on my heart. Loren, my daughter Georgia, and I consulted together about what was going on with me and my ability to continue my business. I had put such intense effort and financial capital forward to create a successful business. Now, I was facing losing all of this. With my present physical and mental capacities, financial loss, and increasing debt, it was decided to let my franchise go. I was not going to be able to build the franchise back, and money was hemorrhaging from my business and my family's financial wealth. Although necessary, the sting of this loss was deeply palpable within me.

Making decisions to move forward in your life is crucial, not just thinking about them and what to do. I had gone from formerly making decisions to having decisions made for me when this all started with the stroke in March. This made me feel angry that I was unable to make decisions on my own or have a say in decisions that affected me. Now, it was time to once again make decisions that affected my life and those around me. There was another franchise that I was aware of, and they were aware of me. We knew about each other, and each had our own clients. I, along with Loren, approached this franchise to see if I could become a part of it. They agreed to bring me on as an associate with the title of Vice President of Sales. I would be offering training and keep doing what I had done in my own franchise. Becoming a part of this franchise would allow me to build a business within an existing one. I was and still am grateful for this opportunity.

The most distressing and depressing moments were seeing everything I had in my office and training center come to the house. Losing my office and training center was profoundly personal for me.

This space was the center of my purpose for my business and clients. Now, this was gone and would not be retrieved or coming back again. Within such upheaval, there were gifts of kindness that would emerge. My lease was up in September, and it was only May. Even though I had to break my lease, the landlord was very kind and refunded the security deposit, which they did not have to do. This helped financially. Some people, including family close to us, were coming to know about our financial situation, which had been evolving, and they did what was possible to offer financial assistance. I was deeply appreciative of their generosity. During this time, another unpleasant decision also arose. It was not clear when I would be able to drive, and we were facing financial hardship. We sold my car, which was financially helpful. It was another forfeiture of my previous life while confronting the current reality of my life. This entire period of time felt like everything had been taken away from me. The persistent and disconcerting thought would invade my mind, "How am I ever going to get things back?" Our house had an overly pervading feeling of scarcity and lack of abundance.

Chapter 5: Getting Back into the Swing of Things

The wedding is now over, and I am slowly getting back into the swing of things. The simplest activities were fatiguing me, and often, I had to retreat early from what I was doing. This included spending time with family and friends and social activities like inviting people over for dinner. My energy was often drained from these activities, and I had to pace myself carefully. Even so, I engaged with family and friends as possible. I enjoyed having my mother stay with us during the wedding. I don't think I appreciated my mother as much as I did during our visit. We really connected during this visit. I realized how much she had done for me, and a heartfelt appreciation flowed through me. I especially valued her point of view. Years ago my mother dealt with a breast mass. I didn't realize then what a serious situation it was and what she went through. She also embarked on a radiation intervention protocol that she found grueling. Through her great and our great fortune, when this was completed, all was well with her. It tremendously helped to talk with her about her past experience.

During this visit together, we also had conversations about her parents, my grandparents, and what they, too, went through during their lives. We were reminiscing about the many stories of the good times my immediate family and her parents shared. I was very close with my grandfather, my mother's father. We endearingly called him "Papa." Papa was a man of character who I idealized my entire life. As a matter of fact, I have wished for my life to be much like his life. He was a generous, kind person with a strong faith and never seemed to be worried about what would happen to him. During his life, he went through 8 life and death surgeries when he was on this earth with us. He never complained and never said, "Why is this happening to me?" My mom and I were talking about when my grandfather was about to have surgery on his brain to remove a benign tumor. I was 17 years old and had breakfast with him down at a Florida house. I said, "Papa, aren't you scared because I know I am?" His very calm voice responded, "Matt, no matter what happens, I know I'm in the Lord's hands." When he said this, I was comforted that everything would be OK. Papa came through the surgery just fine and lived several more years after the surgery. My Papa was my role model in many ways, and I am exceedingly grateful to have had him. I do miss him very much to this day.

As I have previously shared in this book, when someone has a confident feeling of being supported and protected, it creates a deep feeling of reassurance that a health challenge can heal. By keeping this focus and awareness, truly phenomenal healing can and often does occur. An abundance of "vitality or spiritedness or potency" within and around us is always available to tap into and utilize. Each person has their own intricate and intimate connection with this abundance that is often hidden unless we come to know and treat this abundance as a trusted friend, a fierce protector, a loving supporter, a knowing healer, and a wisdom guide. The hidden abundance around us and simultaneously within us can appear uniquely to each person depending on their mind, life conditions, and background of life. Each human being naturally has this occurring. It is truly not an outer experience of living as a human. I realize that I did receive inspiration from people sharing their views like my Papa did. This was an example of the outer experience. However, I also know that I grew to know this inner and outer existing abundance more thoroughly through my evolving meditations and prayers. As this feeling of knowing continued to bloom, I saw and experienced how this manifested in my life situations, especially with this new, unfamiliar health journey that was occurring.

Each person will identify and label this feeling, knowing and experiencing a hidden abundance that can't be diminished or limited around us or within us. For me, the identity and label is God. When I reflected on my Papa and his life, I remember thinking, "Well, this is my turn to be a good role model to my kids, my friends, and anybody who's around me." Through my life experiences with these health troubles, I came to realize that God is bigger than any disease. My grandfather walked the walk fearlessly, approaching his health situation and didn't just simply talk the talk. It was now my turn to go through radiation to eradicate this mass in my brain. I can't tell you how much time I spent thinking about him along with his beautiful wife, my grandmother, who was affectionately known as my delightful Nana. Both were true role models for me, and their lives lovingly consisted of faith and love for each other, their families, and their friends. They believed in the good of human beings and the good of the world. As they knew they had more than others, they sought to be as generous as much as possible.

After my mother's visit, I decided that I would be a better person, including a better father, husband, son, and brother. I dedicated myself to generating and feeling more compassion for people because you never know what people are going through.

With the wedding over and my mother back home, I was going to outpatient physical and occupational therapy. My sister and I more deeply reconnected during this time. Without her, I'm not sure I would still be here today. She also became the catalyst for this book when I told her I would like to embark on writing a book. Her response was, "Why not?" She enthusiastically encouraged me and fully trusted me that I could be the author of my story.

My sister has an Occupational Therapy (OT) background in neurological rehabilitation and specialized in helping people with stroke or brain injury regain motor, vision, cognition, and balance abilities. She fully understood the stroke and the residual difficulties that I was experiencing. She knew that she would need to let the therapists working with me at the University of Maryland Medical Center and Keenan Rehabilitation without her input. She waited until I got home to see what she could contribute to help me further heal. As we live in different states, we did virtual sessions working on OT exercises for moving my left arm and hand and improving my visual skills. Although she no longer practices OT, she still utilizes this knowledge to help people when needed.

Her life has wandered into a unique path of meditation, using the mind for healing and health and incorporating homeopathic remedies. Each of our virtual OT sessions would eventually turn into a lengthy meditation session. Together, we helped to relax my mind and shift thought patterns utilizing the "abundance" around and within me that helped my mind and body heal. These meditation sessions also helped me relax before going to sleep so I could sleep better. Sleep was so important for stabilizing my health and also to help prepare me for the impending radiation that was around the corner. I regularly practiced these meditations alone in the bedroom with noise-canceling headphones on, and I always had good results. In a later chapter, some specifics about as well as accessing this hidden abundance to increase the capacity to heal will be shared.

We both also started talking about homeopathic remedies. She had researched some remedies that I could start taking now before doing any radiation. She felt that we could start to shrink the brain mass now. She had identified several particular homeopathic remedies that we could begin using to deal with the brain mass, as well as a couple of others to help my overall well-being and continued healing. She explained what each remedy was for and sent me research studies on each remedy we would use to deal with the brain mass. The research studies discussed a remedy's innate healing properties and how it interacted with certain brain mass cell types in the brain. These research studies were based on homeopathy and not medical allopathy.

According to my sister, homeopathy is a specific approach to health that harnesses the mind and body's ability to heal. It is based on the principle that a person's mind and body can be influenced to heal by specifically chosen homeopathic remedies written in a book called The Materia Medica. A homeopathic remedy goes through what is called a "proving," in which a chosen group of healthy people takes the specific remedy on a daily basis for a dedicated amount of time. While taking the remedy over time, the healthy person will display "symptoms" which are noted down by the person. When the person stops taking the remedy, the symptoms disappear. At the end of the proving time, all these notes of symptoms and experiences will be compiled into a "remedy picture" within what's called a Materia Medica. Also, notes about what the specific remedy actually is, for example, Hypericum, also known as St. John's Wort, and its healing properties, are mentioned. Matching a person's situation and symptoms with a particular remedy picture follows a "Like with Like" principle. In other words, a remedy is chosen based on the person's presentation of illness symptoms along with a remedy's unique symptoms picture to spur healing and resolution of the illness. This can be likened to lighting a backfire to pull oxygen out of a raging wildfire to weaken it and make it easier to extinguish.

To help me further understand, my sister explained to me that in 1796, a physician, Dr Samuel Hahnemann, in Germany, conceived of and developed the homeopathic remedies system. Homeopathy came to the USA in 1825, and a homeopathic school was opened in 1835. Throughout the 19th century, there were dozens of homeopathic institutions in both Europe and the USA, and homeopathy was considered successful in treating numerous conditions. The presence of homeopathy began to wane at the end of the 1800s, and the last homeopathic school in the USA closed in 1920. Interest in homeopathy arose again in the 1970s. Any alternative healing method is always met with skepticism by the well-lobbied mainstream medical allopathic arena, and homeopathy has experienced this as well. Homeopathy cannot fit easily into the research protocols the medical allopathic path defines as acceptable. Even so, homeopathy does have its own view of its science and does have its validity. The allopathic medical community limits itself by a self-centered belief that "one-size-fits-all" and very often appoints itself with this title.

Between meditating, praying, and taking the homeopathic remedies on a daily schedule, healing of this brain mass had begun. Certainly, by now, anyone reading this book has noticed that "brain mass" has been almost only used, and "brain cancer" is not mentioned except in Chapter 2, which was for terminology. There is a purpose that I am now revealing. The word "cancer" conjures up many emotions and thoughts, almost all that are negative, based on fear, anger, sadness, discontent, and confusion. Words are developed so that we can communicate through language to convey a concept and understand what the concept means. All words have power, and a word's power can be thought of as energy that pervades a person's experience. Some words are so powerful that they seem to have a presence, personality, or even a persona of their own. This is because the word carries an 'experience' that involves many different dimensions.

The word "cancer" is internationally famous, it is internationally known. It has its own fearsome celebrity status. The dimensions that occur around this word are individuals that are given this diagnosis, the family and friends of this individual, various medical disciplines involved with this word, governments, lobbying companies, drug companies, medical technology, hospitals along with healthcare networks, numerous organizations, alternative health options, media, general public interest and gobs of money. In looking at all other existing medical words with all these dimensions just described, this medical word has the most power worldwide. As such, this word has evolved into a potent and formidable personality and persona. In the world of diseases affecting humans, this word is public enemy #1.

The persona of this word congers an image of defiance, like it will do what it wants, dominion over a person's state of health, and a power that is larger than life itself. If it had a form, maybe it would look like some huge monster with a twisted face, bulging eyes, big fangs, sharp claws, and a body made of armor that appears much stronger than us as humans. The personality of this word is depression, scores of fears and doubts, joylessness, anger, sadness, feelings of loss, hopelessness, confusion, lack of power, loss of sense of direction and purpose, diminishing connection to living, life force draining, and life taking. People sometimes say when they are listening to someone stricken with this word, "This is the disease talking, not you." Or "I am seeing the disease right now. This is not you." They are so very correct! They just don't realize the correctness of their observation. As a result, this word can feel like an insurmountable challenge, a surrender to a tough sacrifice or harm and a devastating misfortune that has now befallen. This word is in the driver's seat of fear along with other negativities, and fear is life-taking, never life-affirming.

Looking at the amount of energy that this word consumes is staggering. So many dimensions, as discussed above, are interwoven into the persona and personality of this word. The response that people have when even just hearing this word. Inadvertently, people subconsciously help to generate the belief of the persona and personality of this word with the power of their mind through their thoughts and emotions. Thereby, the persona and personality of the word continue to be further empowered. Our minds are incredibly powerful, although most of the time, we are in the dark about what our minds are and what they are capable of. As a consequence of this word, the persona, the personality, and our minds' identification of this word, it is no surprise that medical interventions are a replica and reflection of this word "cancer."

Like a monster or unknown alien, this word stirs an all-out response, likened to warfare. Bring out all the artillery possible to kill and defeat this, no matter the unintended collateral damage. Movies depicting an enemy that must be defeated and destroyed show these warfare scenes. In trying to defeat or kill the enemy, so much else that is also around is either destroyed or damaged. Ironically, with all this "fight," the enemy is often not subdued or killed if necessary. Somehow, the enemy evades. What ultimately defeats the enemy is a clever tactic, perfectly suited to the situation, that is surprisingly simple and involves minimal or no violence. With this resolution, balance, restoration, and peace have the opportunity to flourish.

Currently, this word, "cancer," has inspired medical intervention methods like the massive, overly prepared artillery response seen in movies to assail this word, this disease. "Less is more" is very rarely considered. However, using more wisdom with less over reactive responses may be a much more viable way to handle this disease.

Yes, of course, this word defines a disease. However, a disease is a severe imbalance within our body that occurs over time and progresses from dysfunction to disease. There are a myriad of reasons for this to happen, and we are multi-faceted human beings. We can unlock multi-faceted knowledge that deepens our understanding and wisdom about this disease process, along with diverse approaches to effectively address it. In this context, a "one size fits all" approach is entirely inadequate, and a broader set of skills is necessary to treat this health situation appropriately.

My sister refuses to use this word. In simplest terms, cells are dividing in ways they are not supposed to be doing. They are acting inappropriately and creating some havoc. So, my sister calls them "renegade cells." Renegade means to desert or betray and become a traitor. These cells are doing this; for whatever reason, they desert the correct way to be, betray other cells and the body, and become traitors to the correctly functioning cells and body systems. Defining this way creates a clear visual of unfriendly, unwelcome guests who have come to sit the body and now need to be dismantled, eradicated, and completely banished. Where there has been disruption or damage, as much as possible, restoration must occur. So, my sister and I only discussed the renegade cells and how we were going to handle them and take charge of them. Sometimes, such renegade cells may not be fully hostile, and their process may change to stop harm. Most of the time, the renegade cells are hostile and require banishment, eradication, and a change of conditions so that other hostile guests do not reappear. In this situation, however, the power of clear wisdom and proper compassion must supremely be in command.

As I have said, words can be harmful even if this is not the intention and people are well-meaning. It seems people, including medical professionals, are confused, thinking that this word is now what I am, which I am not. My sister would tell me not to "inherit" this word and the disease it refers to. From talking to each other, I became emboldened to see it as something that was happening, and rather than fully participating in it, to observe it instead. In other words, she advised me not to claim these "renegade cells" as part of me or mine because there is always a natural state of health possible within us. Our body is constantly engaged in a process called homeostasis. This amazing process is our body's way of maintaining stability and balance while adjusting to changing conditions. It's not just our physical functions that are being regulated—our emotions and thoughts are also an integral part of this alignment and balance.

When someone said to me, "You have cancer," a deep, negative, heavy feeling cut through me physically, mentally, and emotionally. It was like a 1000-pound weight crashed onto me. I remember this vividly. "Having" implies, "This is yours." When I heard these words, it was like this was now an astronomically unwanted part of me that I had no say if I wanted or not. Everyone around me now wanted me to acknowledge this as now being mine, as this is now me, as this is now my life, and as this is now my identification with myself and others. I did not feel this way and did not want to do what they said. In thinking this way, I was outnumbered at the time by everyone else around me. This was extremely isolating and overwhelming, and I felt lonely.

When my sister and I could finally talk over the phone, she validated my thoughts. While talking, she and I fixated on the fact that these renegade cells had somehow bypassed my mind and body's natural defenses and had taken up residence in my brain. However, they weren't supposed to be there, were absolutely unwelcome, and would be kicked out—or, if necessary, fully eradicated. It felt like we were speaking directly to the renegade cells rather than me. During our conversation, we determined that saying such words as "You have cancer" was extremely damaging to healing, even dangerous to the person due to the dimensions, persona, and personality of this word, so we discarded using these words. If necessary to use this word, she suggested something like, "Cancer has appeared within you" or "Cancer has taken up a space inside of you" could be more helpful because I or any other person would have more ability to separate himself or herself from an unnatural condition that truly does not belong to them. In such a way, better possibilities for healing and health could surface.

This understanding of separating ourselves from a disease is essential for healing that might not otherwise be possible. As discussed earlier in this chapter, this 'abundance' within and around us reveals and manifests uniquely for each person. For me, it's almighty God. For my father, it was "his angels." My father always said he could feel his angels around him helping him and caring for him. He talked about them quite often and had confidence that they were always there, never leaving him.

My father and I have eerily similar parallels with regard to our health path. Like me, my father went through some serious health challenges, one being a critical heart attack. After he returned home from the cardiac critical care unit in the hospital, he began having fluid in his lungs, which can be a common side effect of a heart attack and bypass surgery. Somehow, his lungs would not clear of the fluid and, after 1 month, worsened. He went to see his primary care doctor, who did a chest x-ray and notified him that he had pneumonia and would be starting oral antibiotics right away. She also told him that if the pneumonia did not clear within 5 days or so, he would be admitted to the hospital to start IV antibiotics.

My father, a bit shocked, got in the car and started thinking about this pneumonia and was thinking, "What are you doing here? You shouldn't be here." He kept thinking this quite obsessively and, by the time he got home, was now "talking" to the pneumonia. He was telling it that it was a guest in his body that he did not invite and was telling it to "get out now. You don't belong here. You are a terrible guest!" Pretty much, he was incessantly talking out loud to the pneumonia all day.

That night, around 1:00 am in the morning, he was awoken by a deep red light emanating from his chest. As he was lying in bed, he saw these small white-colored beings excitedly fluttering above his chest when he became more awake. Large balls of light would grow in their hands, and then they would throw these red balls of light to his chest. When these balls hit his chest, his chest turned a glowing crimson red. Immediately, he knew he was being healed from the pneumonia. Excitingly, he woke my mother up, "Gloria! Do you see what is happening? Do you see them? Look at my chest; it's glowing deep red!" My mom, waking up from the stupor of sleep, couldn't see anything he was relaying to her. Panicked, she rushed out to bed to get the BP cuff and started to call 911. She thought he was dying. My father was following her around the house, saying, "No! No! I am not dying! I am being healed from the pneumonia! My angels are healing me! Don't call 911!" Although she could not see anything, she eventually believed him. His experience with this phenomenon lasted 45 minutes to 1 hour. When "his angels" and the crimson-red color disappeared, he knew his lungs were healthy.

The next morning, he called his doctor to announce he no longer had pneumonia and would not be taking antibiotics. Alarmed, the doctor demanded he come in that day for another follow-up chest X-ray and a firm discussion about why he would be taking these antibiotics. He agreed to come into the doctor's office for a repeat chest X-ray. The chest X-ray came back completely clear, and my father started to share this incredible experience he had been through the night before. Rather than being interested, she quickly shut the conversation down and ignored what he was trying to share with her. Instead, she said that maybe she had read the initial X-ray incorrectly, even though the radiology report showed pneumonia. My father knew without a doubt what happened. He experienced his hidden abundance of health in his distinctive way.

After discussing remedies, understanding what they do, and seeing that they would not bring any unwanted problems to me or my condition, I trusted with my heart to begin taking them. I was on a daily schedule and did feel something after I took the first remedy. I felt clearer and more energetic. Some people may say this was a placebo effect. People can decide what they wish. I just know that I felt lighter, and my brain felt clearer.

My 1st MRI was done in March when I was hospitalized, and I had a 2nd follow-up MRI in May before starting the radiation. I had been taking the homeopathic remedies for 2.5 weeks before the May MRI. When I saw the result of the follow-up MRI, I texted my sister to email her the MRI report. I was surprised by what I read and wanted her to have her take on the MRI report. The MRI result said, "The mass overall has decreased in size since the prior study," and documented the change in size measurements between the March and May MRIs. The decreased size was significant, and I was excited about this change happening before doing any medical intervention. I was also curious, thinking, "Was radiation even necessary if the mass is already decreasing before starting radiation?" My sister agreed that this finding should be examined more closely and carefully. Maybe continue the remedies for another month and then get another MRI to see if the brain renegade cells were decreasing.

When I shared with Loren that the renegade cells had decreased in size, understandably, she did not know what to think. It did not seem possible to her that the renegade cells in the brain had decreased. On the other hand, both my brother and sister felt this was surely possible and really wanted me to push the doctors more to slow down and look at this result before heading onwards with radiation. I was uncomfortable sharing with the doctors that I was taking homeopathic remedies. My sister did understand my predicament and shared my concerns about revealing the remedies to the doctors. She was familiar with many instances of medical professionals being negatively judgmental, intolerant of, or flatly rejecting healing methods other than their own.

In retrospect, I should not have kept the remedies a secret from the doctors, but I did not feel comfortable sharing them because I thought it would lead to more questions than I had answers to. I was also keenly aware that they were adamant about their medical intervention protocols. If I had received a lot of pushback from them, maybe this could have affected my confidence in my choice to take the remedies. In the worst case, I feared that they would forbid me to take them or force a choice between themselves and the remedies. These possible scenarios in my mind created the decision to withhold from the doctors that I was taking homeopathic remedies. Within my heart, though, I felt the remedies were working and helping to decrease the renegade cell mass.

Chapter 6: Confronting Death: Minutes to Seconds

The first day of radiation was approaching. The first visit before radiation was fitting a face and head mask. It was the end of May, and my youngest daughter, Lucy, was creative and designed a "radiation sessions chain" made of red, white, and blue cardboard paper. The original plan was for the radiation to be finished by July 4th. She did a similar chain when I was going through dialysis. My wife and children supported me every step of the way, staying as positive as possible while traveling this path. Every time I came home from a radiation session, I would tear a link off of the chain – another session was finished, and now there was less to go. It was a special message that was motivating to me. It is so important for the people around someone who is going through health challenges to include messages of hope. It's very valuable to have something to look forward to with the reminder that a person's current situation is temporary. This reminder opens the door to look through that reveals a greater abundance waiting for them in the future.

I really felt that the meditations, prayers, and homeopathic remedies were already healing my brain and that the renegade cells were decreasing and being eradicated. I had been feeling better than before. The MRI also showed a significant decrease in the size of the renegade cell mass. However, I didn't see any way to share my view of what I felt was going on, and it still didn't feel safe to tell the doctors what I had been doing. Understandably, my wife and children were skeptical that my current approach was working to heal this health condition. I really didn't have a way to discuss the MRI or the other methods I was doing with the doctors before the 1st radiation session. When Loren and I arrived for the 1st radiation session, we had a consult with the doctor overseeing the radiation. He told me that the decrease in the size of the renegade cell mass was due to cyst rupturing, and it may be that cysts could develop with the renegade cell mass. While this could possibly be an explanation, I wasn't really convinced. However, I was keenly aware that my wife and children were suffering from this health condition I had, were afraid, and just wanted this thing out of my head for me to be healthy. The doctors were also confident about their plan. Although I did have some doubts, I still pushed ahead and started the 1st radiation session.

We had about an hour's commute to the proton center where the radiation sessions took place. During the ride, there was not much communication between Loren and me, mainly because it took a lot of energy for me to strike up a conversation. Loren kindly created a playlist of Christian songs that we listen to. During the ride, I sat silently, enjoying the music. From my experience, I cannot overstate the importance of being buoyed and surrounded consistently by music, TV programs, social media, personal conversations, or anything positive and beneficial. Having this atmosphere enveloping the person truly offers sustenance and nourishment for strength during the rigors of a health dilemma, along with healing, which was vital for me. How this looks for each person going through a health challenge is unique to their characteristics.

When we arrived at the proton center, I was pleasantly surprised by how nice and inviting the physical environment was. The staff was welcoming and ensured we had everything needed for the radiation session. Even with all of this, I found this visit for the 1st radiation session intimidating. It was a lot to take in. When they called my name to receive the radiation, we walked back into the area where the session would take place, and I felt like I was walking into a spaceship. The radiation technicians did a very nice job ensuring my set-up and that I was as comfortable as possible during the radiation. They also ensured the music playing in the background was something I wanted to hear, which was a nice touch. So, I got settled in, and before the radiation began, the technicians brought in a very tight-fitted mask to wear over my face and head. On the mask, they had markings, or what they call tattoos, of the area where the radiation was going to target. I simply laid back with my eyes closed after putting the mask on my head. I pictured beach scenes in my head and prayed throughout the entire session.

After completing my treatment, a couple of technicians came to remove my mask and help me get up from the table. Upon sitting up, I was very dizzy from the radiation. The technicians told me to get up and walk to the waiting room to see my wife. When I got to her, we walked hand in hand down a long hallway to an elevator that took us to our car. Our car was waiting for us by the valet service at the proton center. As I mentioned, the proton center where the radiation sessions were taking place was physically a very nice place, and the staff, including the valets, were wonderful and did their best to make us as comfortable as possible.

As Loren and I drove home, it was early in the afternoon, and both of us were excited to have 1 session completed. Once again, on the way back home, I was silent in the car and did not converse much. This was not because I did not desire to, it was instead that my energy levels were just drained. Upon arriving home, tearing off my first link on the radiation chain felt really good. I thought to myself, "We're going to get through this," and I fully expected that the mass would be eradicated.

The next day came with the same routine as yesterday. I was doing my best to block out any outside interference or noise from the TV or even from the conversations around me. I was focused again on getting the radiation done and looking forward not only to ripping the next link off the chain when I got home. I eagerly wanted to finish the entire chain day by day. Not only this, but I also visualized the day when I could "ring the bell" at the proton center after completing the last radiation session. What I learned from going into radiation was that each time I would finish the session and get off the table with the technicians, I was going to be dizzy, and my energy level was going to be very low. It was such a joy having my lovely wife there with me and seeing her face each time I completed a radiation session.

During the radiation, I was still taking the homeopathic remedies and praying hard during every radiation session. I didn't realize how concerned my sister was about the radiation. She never shared the disconcerting feelings that kept pervading her mind about the radiation. She was very aware that telling me what her mind was experiencing and seeing would certainly not help me or my family. No matter how much she meditated on me or thought about me positively and "sent" me healing energy, she was unable to shake that something serious was going to happen. She had an overwhelming sense that my life could be in danger. She felt the radiation would be too powerful and too much and very possibly cause irrevocable injury to me.

She did not confide in my mom either because my mom was doing everything to remain positive and support me during all of this. However, she did confide in my brother, Michael, what kept arising in her mind. It turns out that Michael was similarly immensely uncomfortable and disturbed when I was starting radiation sessions. Both of them felt that there was not enough recovery time in my brain from the previous stroke and that my brain was vulnerable from having a stroke, a biopsy very soon after, and now radiation just a couple of months after the previous hospitalization and rehabilitation. My sister also felt that the doctors were not doing "due diligence" to understand what the strokes were related to. She felt there was an underlying situation, probably with the brain blood vessels, that should be more definitively addressed before embarking on such an invasive medical intervention. She knew the doctors would just say that these were occurring from the renegade cells. The renegade cells were the "blame." And yes, this is what they believed—however, no proof existed of this hypothesis. There was a stroke back in September 2019 when the kidneys went into complete failure and one in March 2023, and a long history of kidney disease that she felt was not being taken into account before doing radiation to my brain.

My sister felt a lot of personal pressure to protect me from the effects of radiation. She was very cognizant of the myriad of possible harmful side effects that could and often do occur, although she did not share this with me. It was her deep desire for me to be confident, calm, and positive because she knew the power of my thoughts would help steer the outcome of my health. Like everyone else, she wanted a stellar result for my health and wanted me to be fully restored and have a long, happy life. She was not just concerned about the duration and number of sessions and the amount of radiation I was being exposed to. She was equally feeling tense about the fallout of effects following having radiation weeks, months and years after. My papa, who I talked about earlier, had radiation to his brain after the regrowth of a benign tumor called a meningioma, as he was not a candidate for surgery due to his heart. She saw firsthand what radiation had done to his function during radiation and subsequently for a long duration of time afterward.

She switched the homeopathic remedy schedule to not just what we were doing for the renegade cell mass. She also added more remedies to help mitigate the effects of the radiation on my brain, my body, and my functioning mind, as well as to try to protect the integrity of the blood vessels in my brain. Each day, I would receive a text of the daily remedy schedule. Even with all she was trying to do, she still had the feeling that we were on very shaky ground doing radiation, especially five days a week for continuously five weeks. It unnerved her each time when the thought came into her mind of a stroke occurring. I did not know every day I had a radiation session, she did a mediation session of protection and healing mantras, visualizing me receiving them and deeply holding aspirations that I would be ok and get through each session as strong as possible with my health being safe.

I kept everything as positive as possible. Even though my energy was mostly low, my concentration and thinking diminished, and there was more and more loss of appetite for food. I always thought that I was doing well with this and was ok. During the 2nd week of radiation, on Friday, by the time Loren and I arrived home, I was unable to speak properly and was having problems finding the words. After some time, this went away. Loren immediately called the proton center to report what had happened. The proton center reassured her that this was nothing to be concerned about.

My brother, Michael, found out the next day when I told him what happened. He had come over to visit me. After hearing this, he was very apprehensive and upset about the radiation and undoubtedly now had misgivings about my health and safety getting radiation. He called my sister after learning this, and she was instantly disturbed and could not believe that the proton center had told Loren that this was not concerning. With the history of a previous stroke just in March and a previously long-term kidney disease that also can affect blood vessel integrity, she was flabbergasted at this seemingly non-sensical response from the proton center. She wanted a CAT scan or MRI done pronto. The renegade cell doctors had advised Loren and me that a stroke could be a possibility with radiation to the brain. Radiation can affect the blood vessels around the area being radiated, and the consistency of blood can also be affected. However, they felt they had a good plan to address the renegade cells, and they convinced us to proceed forward.

My sister wished to voice her concern about my word-finding problems after the radiation and adamant feelings to stop radiation, get a scan done and not resume radiation until it was clear what happened. However, she was aware that Loren and I had committed to the plan for radiation. In her experience, doctors who deal with renegade cells often seem undeterred and gung-ho to proceed onwards with their interventions without looking at the larger picture of what is going on with the person's health undergoing these interventions. She has seen the balance of so-called power between doctors, which also seems imbalanced between renegade cell doctors and other doctors involved with other medical care decisions. Renegade cell doctors seem to be more in control of a shared patient's medical care and decisions than other doctors. She had observed this phenomenon with clients she has worked with as well as with our father and the Multiple Myeloma condition he was experiencing.

When my father was emergently hospitalized in full acute kidney failure, a physician assistant (PA) from the oncology group that was treating him entered the hospital room. With my mother standing next to my sister and my father in a very sickly condition, laying in a hospital bed on the verge of sepsis, the PA announced to my father that they ordered an MRI with contrast to check the status of tumors on his spine and possibly in his intestines. All my sister could do at that moment was stare at the PA and try to hold it together. My father's urine was dark brown color from the sudden acute kidney failure. He was almost in sepsis that the hospitalist doctor was doing everything to prevent, had blood in his stool sample due to damage of the intestinal wall from a blockage, was still in severe pain, and a blood transfusion was being planned in the next couple hours because his hemoglobin was so low.

Functioning on no sleep for over 24 hours, she incredulously asked, "You do realize that the contrast has to go through his kidneys to excrete, and he is in serious acute kidney failure?" as she pointed to the catheter bag with dark brown urine in it. She was looking at my father's face of disbelief for having an MRI amid everything going on with him. The PA, responded, "Yes, but we need to know what is going on with the tumors." My sister replied, "At this moment, the only thing we care about is my father getting medically stabilized and out of the extreme pain he is in. He is in no condition for an MRI with contrast, and right now, we don't care about the status of "the tumors." All we care about is him, his life, and his current medical condition. Please leave. We will talk to the hospitalist doctor."

The PA did leave and my sister went directly to the nursing station to have them call the hospitalist doctor immediately. Within a short time, the hospitalist doctor called and spoke to my father, mother, and sister. "Don't worry. We are not doing any MRIs at this time. Yes, the kidneys are very weak. I am making sure this is clearly communicated and will come to see all of you a little later." My sister was very relieved; however, also really troubled that if she had not been there, the oncology medical group through the PA may have overruled the hospitalist doctor's medical care planning.

I also had a neurologist regarding the stroke who was involved in my medical situation. However, any time I asked a question regarding the stroke or the renegade cell mass, the response was to refer this question to the renegade cell doctors. It was like the neurologist had nothing to say or very little say in my medical care plan. I do admit that even though the renegade cell doctors were kind to me, I often felt they held all the cards, and I had to follow whatever they told me, even if I had doubts. On a very personal level, my sister witnessed this phenomenon with our father. Even though this had been a few years earlier, the memories were still fresh.

Fear was also living very strongly within my immediate family and my mother regarding these renegade cells. It seems that families, through their fear, unintentionally infuse this fear into their loved ones experiencing a renegade cell problem. Everything is fearful, so there's a feeling of pressure to conform to what renegade cell doctors are saying to do. Of course, the family certainly wants their loved one to be completely better and move on with life. As a loved one with this problem, I desire this as well. Therefore, it was full steam ahead even though there had been a clear warning sign to proceed much more carefully with caution.

On June 20th, it struck again. It was not what we were counting on. Another ride in the ambulance to Carroll Hospital in the morning. My left side went completely limp with no mobility or strength. Unbelievingly and unfortunately, it was another stroke! By the time I arrived at the hospital to the ER, my left side was restored, and the CT scan showed no signs of a stroke at the time. Several hours later, however, an MRI showed two more strokes, a smaller one from a couple of weeks ago when my speech became slurred right after the radiation session and then one from this morning. Such disbelief. Such sadness. It was an immense tough blow because we all in my immediate family thought we were making real progress with the radiation sessions shrinking this mass in my brain.

Truly, everyone within my entire family and myself were heartbroken. The renegade cell doctors told me that the radiation would take care of the strokes. Nothing could be further from the truth. The left side of my body was also weaker than I originally thought. I was going to be admitted into the ER overnight for observation. I texted both my brother and sister, "had another stroke." Their hearts were heavy, and each of them was profoundly affected that this was happening to me. What they both felt could happen now had. What my sister had been feeling day after day now was reality. For her, a nightmare come true. Both felt my situation was spiraling out of control, and it wasn't clear where things were going to land. They were frustrated and powerless to change the course of the direction that was occurring. My mother was in shock that they were all going through this with me again. She was praying I would be okay, although she was worried at the same time.

During this 1st day at the hospital, I had recovered many functions from the day before, and the overnight doctor who visited me said it was such a miracle that so much recovery had already occurred. I was able to speak clearly to him and remembered everything that had happened. He said, "God's grace is shining upon you that you can speak and remember. It's truly amazing." He was very pleased. I asked the doctor if radiation could have caused this stroke. He said no. Even with what he did say, it really could not be ignored that I had word-finding and speech problems on the way home from a radiation session, and the MRI showed a stroke that corresponded right to this time frame. He said that on the MRI, the renegade cell mass was clearly much smaller. I wasn't sure what that meant, but I was happy. Both PT and OT had done an evaluation and mostly just saw balance and coordination issues. I was walking okay, and my strength and speech were good. All of this gave me great hope that this episode would be over very soon and I could get back home.

The next day, the opposite happened. Since I was so mobile, I was allowed to get up and go to the bathroom, except this time, I struggled with balance and couldn't walk in a straight line. I was leaning to my left and ended up leaning into the wall with my left shoulder pinned to the wall as I was walking down the hallway. I just couldn't control how I was walking and moving. I was uncoordinated and unstable. I was thinking, "If anyone is watching, they are wondering… why am I walking in such a way to the bathroom?" I was essentially hugging the wall and walking down the hallway. The nurses saw me and came to help me back to my bed. The day before was a miraculous recovery, but today, it is now a long road ahead. Loren had been at the hospital earlier, and Michael was on his way to visit and see me. When he arrived, he instantly saw that I was not in good shape. He noticed that I was looking at my left arm and leg and seemed to concentrate on making them move. I was not able to move my left leg or arm well. This was a red flag. He also saw that my hands were completely white in color, including my fingernails. Without any medical knowledge, he really did not know what to do and was trusting that the doctors or nurses were monitoring me.

He had to leave the hospital and really did not want to go because there was a nagging feeling inside that all was not well with me. Later, after everything happened, he felt he should have been more pushy with the doctors about what he was seeing. When he got to his car, he called my sister, who also knew none of this was a good sign. She felt nauseated and sorrowful, thinking everything was possibly going downhill and there was nothing she could do about it. In reality, the stroke was extending while I was lying in bed. At some point, the doctors became concerned about my lack of ability to move my left arm and leg and ordered a stat MRI—very bad news. I had a blood clot in my brain. I had been watching the faces of Loren, who had arrived back at the hospital, and all the hospital staff change from being happy with good news to worry. The good news face was gone while the worry face remained. I was praying very hard that all of this was going to go away.

I was moved to another room. There was suddenly a flurry of medical staff rushing around me. I heard the doctors telling Loren I needed to be flown by helicopter to Sinai Hospital. They were sending me to the Interventional Radiology Unit. Loren called my mom and said to her that her heart sank when the doctor told her this news. Hearing all this news made both Loren and me very afraid. I needed a medical procedure called a Thrombectomy to remove the blood clot in my brain. A female minister was nearby and saw how frightened both of us were with everything going on. She came over to us and asked me, "are you scared?" I responded, "I am very scared." She right away started praying with us. While we were praying, we learned that the weather was too bad for a helicopter flight to the other hospital. Instead, I would have to be sent by ambulance. EMT and medical staff were now quickly scurrying around, getting an ambulance ready. Time was of the essence, and time was promptly ticking away. I was rushed out on a cot outside to a waiting ambulance. It typically was a 30-35 minute drive to the other hospital. I was fully aware of the seriousness of what was happening to me. The medical staff, EMTs, and now ambulance were moving so fast to get me to that other hospital by any means possible.

While on the cot, I was strapped into the back of the ambulance with an IV hooked up to me, and off we went. I am not sure how fast we were going, but we made it to the other hospital in record time. Once we arrived, the waiting medical team was wasting no time. Things were moving very rapidly, and I was moved from the ambulance right into the operating room. I was hoping I was going to have anesthesia to put me asleep for this. The 1st doctor to attend to me, the anesthesiologist, notified me that I would not be receiving any anesthesia or be put under in any way. The OR table that I was lying on was cold and uncomfortable as it was basically a tile table.

Time was ticking, and the window to save my life was getting smaller and smaller. A wire was being shoved through a blood vessel in my thigh, the groin area, on the way to my brain to the blood clot. I am not sure how many medical staff were in the room; however, there were a bunch of them, and all of them were working feverishly to remove this blood clot. The wire was being pushed through my body through blood vessels to get to my brain. The wire needed to get to my brain to scrape off the blood clot and restore blood flow to my brain. I am not sure how long the procedure lasted; however, I felt the wire going through my body and up to my brain as I was wearing headphones. I became conscious that I was in a very critical condition. I felt when the wire reached my brain, and it started to scrape the blood clot, which was excruciatingly painful. I still had my eyes shut and could hear the wire scraping the blood clot. For some reason, my ears became very, very hot, and I heard a ping. Deep inside of myself, I was praying that this guy doing the thrombectomy was going to get me through this when suddenly the doctor asked me to move my left leg, which I did. The entire room of medical staff broke out into clapping and cheering. I was in medical shock. However, I do think I remember the doctor getting down on the floor and doing pushups as a celebration after I moved my left leg.

A wave of relief washed over me that the procedure was finished and successful. Yet again, I acknowledged I was fully spiritually protected and also had the exact right people to do this Thrombectomy that just saved my life. I felt the loving presence of my God with me, and then, as well as now, I feel that I am living proof of the grace that is God. As I was being wheeled out of the operating room, I was thinking to myself, I have a new life ahead of me, and I want to make it the best life possible. I was still here, present in our world. I was minutes and seconds away from not being here, yet here I was. The best life possible to me means helping others who are facing challenges in life. Assist them in knowing and experiencing their hidden abundance and knowing that this abundance supports them, helps them, protects them, guides them, and heals them. I wish each person to know, feel, and have this experience. For me, it is the knowing, feeling and experience of God. I was on the way to a stay in the ICU to have round-the-clock direct nursing care.

As mentioned before, me and my father have shared strikingly similar episodes with dramatic health events. When his heart attack occurred, he was golfing and came back to the pro shop, not feeling well. Like me, he was loaded into an ambulance to a local hospital, and his medical condition rapidly deteriorated in the local hospital. A helicopter was urgently ordered to get him to a specialty hospital, in this case for the heart, 40 minutes away. In his case, the weather was good, and he was flown by helicopter under very life-threatening conditions. When my father was being loaded into the helicopter, one of the flight nurses turned around to the ER doctor, watching in the distance. Not realizing my mom was nearby and saw him, he turned both of his thumbs down to the ER doctor, a negative sign that my father would live. Like me, the medical team was waiting for him as he was speedily whisked from the helicopter pad to the heart catheter lab with only minutes to spare. Like me, there was no time for anesthesia while the catheter wires were swiftly being pushed into a blood vessel in his groin to get to his heart. Like me, his human life was minutes to seconds away from no more existence. The image of the heart showed a trickle of blood flow to the left side, and no blood flow on the right side of the heart. The doctors were yelling at him to stay still and not move, although he was snoring and completely not responding to them. Like me, the block to the blood flow was critical and in a very bad place. For him, the 90% heart blockage was in the "widow maker," a major blood pipeline to the heart. Like me, when they were able to clear the blockage to the blood flow and insert a temporary stint, they cheered. Unlike me, my father remained snoring and was unaware of the gravity of the procedure or when it was over. Although he was awake in the helicopter, by the time they started the heart catheterization, he was snoring. Like me, after the procedure was done, he was admitted to an ICU unit, in this case, the cardiac critical care unit for intensive nursing and medical care.

I don't know the chances that a family would have two family members in a health crisis at different times resembling so many intricate details. When I was in ICU, my sister and brother discussed this with each other and marveled at the similarities between my present situation and his previous one. I think why both of us shared such an amazingly comparable experience may always remain a mystery. It's known, though, that besides having such a path together, we each knew this hidden abundance that assisted. After his experience, my father talked about feeling the abundance of his angels when everything was going on.

Chapter 7: The Aftermath

As of now, I was feeling the presence of my Lord, my almighty protector, around me again as the medical staff was cleaning up all over me and moving me to a cot to wheel me to the ICU. I was looking forward to seeing my family, closing my eyes, and getting some sleep. I was so in awe and grateful to everyone who helped me get through these precariously precious moments, this emergency procedure. I was thinking to myself about the future I was going to have spending time with my future grandchildren and how this experience has changed me forever. Inwardly, I was expressing thankfulness for the surgeons' expertise from the first having a brain biopsy to now, this time of the essence of life-saving intervention. I expressed appreciation to God for blessing the medical staff's intelligence, skilled hands, and professional equipment to get me through this latest round. I became very optimistic that I can help others who face health challenges and other challenges make it through by having an abundant mindset, trust, and the right people around them so they can prosper.

All of this was running through my mind while being wheeled into the ICU. I did not know what to expect in the ICU and was relieved that I headed into a private room. I was receiving 24-hour nursing care, which meant not getting much sleep. I was allowed to sleep for 15 minutes at a time before being woken up to check my vitals. At night, I had a head nurse along with three other nurses who assisted the head nurse. Not only would they wake me up every 15 minutes, but they also hooked me up to many monitors and checked them throughout the night. Around 2 am the first night, the transport person came to get me and wheeled me down to the MRI machine along with a couple of nurses. This trip to the MRI machine was much easier than in the past because at 2 am, the area was not busy, and they could see me right away. The scan was done quickly, and I was placed back on the gurney to be wheeled back to my room. I took this as good news because I was going back to my room. I found out shortly after that the blood clot had been successfully removed. Within myself, I thanked God, along with my loving Jesus, for the news. And once again, I mentally thanked the surgical team. I looked forward to getting some sleep and was excited to see my family in the morning.

I woke up at 7 am the following day and was excited because my brother Michael was coming to see me. Michael called on my cell phone and mentioned he was not able to get to the hospital because he was travelling from New Jersey and would not be able to make it to the hospital. I was disappointed to hear this news, but I understood because he was splitting his time between New Jersey and Maryland. Michael and I had grown apart in our adult years. He experienced his own challenges over the years, and now I appreciate being closer to him, which is similar to when we were kids. Within all these circumstances, Michael was now one of my biggest supporters despite what occurred over the last few years.

My wife Loren came to visit as she always had done. I was eating my breakfast when I learned that the physical therapist was coming to my room to take me for a walk. This was going to be an evaluation to help determine if I could be transferred from the ICU to a regular room in the hospital. I was up to the challenge. I have to mention that the young man who took my order for breakfast and brought it to me went above and beyond to make sure he brought food I would eat. I was hungry for yogurt and fresh berries. It's hard to believe the hospital did not have any berries. This young man went out of his way to find a large plate of fresh fruit that he would bring me every morning in my ICU room to replace the yogurt and berries. The fruit was beautiful and fresh, and it got a lot of attention from the nurses and others who came to my room wondering how I was able to get such a delicious fruit platter. I had a lot of pleasure eating such amazing fruit plates in the morning.

This morning, the nurses had moved me to a chair so I could sit up and eat my breakfast while Physical Therapy was on the way to get me to have me move around. As I mentioned, I was ready to take on this challenge because if their evaluation went well, it was one step to being discharged from the ICU to the regular hospital and then one more step to going home. My human guardian angel, Loren, walked into my room while I was finishing up eating the fruit. She was pleased to see me sitting in a chair having a healthy breakfast. The nurses gave her a rundown of the day's events, including going for a walk with the physical therapists. The two Physical Therapists walked into my room and asked, "are you ready to go for a walk?" I said, "Yes," and they rushed around to get me a walker. They laid out the path we were going to take by walking around the entire hospital ICU floor.

When we started walking with the walker in the ICU hallway, I said, "I don't need the walker," which was an astonishing blessing considering I could not move my left side less than 48 hours earlier. Reluctantly, the therapists agreed to let me walk without the walker. They presented me with visual targets to walk towards without bumping into people in the hallway. In addition, they gave me a provocative task to count backwards from the number 100 by 7s while walking. I remember thinking while also concentrating on walking "100, 93, 86, 79," etc. This was not a very easy endeavor because of dyslexia and not thinking clearly from the recent strokes while walking at the same time. I made a deal with the therapists and said, "If I'm doing this, you two have to do it as well." The funny time came when we were all counting out loud, and suddenly, both therapists walked into a couple of trash cans in the hallway. This comic relief made me feel better about myself in the way that I did not have to be perfect with every exercise I was trying to do.

After walking around the ICU floor without any assistance, we made it back to my room. The Physical Therapists gave a glowing report to the nursing staff. The nursing staff now had the confidence to allow me to walk to the restroom located in my room when needed. I was pleased that the nurses shared all of the good news of me being medically stable and able to walk around the ICU floor without any assistance when the doctors came in to check on me that morning. They dished out to them everything that I had accomplished during the morning. On the agenda, I had one more test to pass. A swallowing test includes swallowing chunks of food, drinking from a glass or a cup, and drinking with a straw. Fortunately, I had no problem swallowing and passed with flying colors.

By the time this was completed, it was now mid-afternoon. The fantastic news came: I was being transferred to the main hospital as long as a room was available.

With great fortune, we did not have to wait too long to learn a room opened up one floor below the ICU. The nursing staff was terrific while rushing around, gathering my belongings, and getting me cleaned up to be moved. Leaving the nursing staff after three days in the ICU was bitter-sweet because we developed a comfortable routine, and I felt safe there. Now, I was about to leave the ICU, where 4 nurses were taking care of me at a time, to having just 1 nurse. I didn't know this nurse and had some trepidation. However, these nurses reassured me that the nurse in my new room was very competent and fully able to provide good care to me.

When I arrived on the floor of the main hospital below the ICU, the atmosphere was so amazingly different. I briefly met my nurse on the way to my room. After arriving at my new room, the nurse came to check on me. She made sure that I felt at ease being here in this hospital room and made sure I knew all the buttons to push if needed: The nursing call bell, TV remote button, and emergency button. Then, she was off to her next task. Suddenly, after all the hustle and bustle, when I first arrived by ambulance at Carroll Hospital ER to now this hospital in a full-blown emergency condition, then under constant surveillance in the ICU, I was alone. No one was around except myself.

For the first time, there was a quiet space to contemplate all that had just occurred over the past several days. Like a lightning flash, it became very clear to me that I was going to write a book about all these experiences I had been through with the goal of helping others who have or are going through similar challenges. I ordered dinner and decided to turn on the TV, which was unusual for me because I don't watch much TV. As I turned on the TV, I instantly met with "the news" on the channel the TV turned on. As usual, the news was about something negative that had happened. At once, I felt the intensity of this news and completely turned off the TV because I immediately noticed how this news channel affected my mind in a bad way.

In my view, most news seems to be "junky," just reporting nonsense occurrences that should not really receive any attention or over-accentuating occurrences where harm or some other drama happened. I don't find either one of these approaches to "reporting the news" helpful, and in no way was this helping my process of getting better and healthier. I have come across so many great people from unexpectedly embarking on this health odyssey. Therefore, the last thing I wanted to watch was the news, which was nothing but bad news, from the pandemic to the political circus that was going on to natural disasters. I'm not saying everyone should not be informed. Still, with so many good, hard-working people in this world who are not recognized or rarely are, I have found that the news often being reported is not actually happening in the real world and is inaccurate. So, TV stayed off and was replaced with the priorities of relaxing my mind, respite, and healing my body.

My sister was able to be in contact by a phone call on my cellphone with me and Loren when she heard the news that I was being moved into the regular hospital. She clarified with Loren what homeopathic remedies would be for assisting the recovery from everything that had just gone on, as I wanted to start taking the remedies again. She also described what each remedy was and what it was for so that Loren and I would understand. We would also continue the remedies for the renegade cells in the brain. Loren brought these homeopathic remedies to the hospital. A lot was continuing to go on, and we still had to deal with this mass in my head. Loren went home when I switched rooms to where I was now. She came back to the hospital to visit me and see me. She brought a small pouch of remedies, a toothbrush, and an electric razor.

It may sound quite humorous, but I was ecstatic with the razor and toothbrush because I knew I was getting better now and could get up to brush my teeth and shave. Taking a chance on the TV once again, I found a PGA golf event that was perfect for me to watch. My dinner arrived on a tray, and it was an ideal time to talk to my sister about the remedies she shared with Loren to bring and which ones I should take, so I reached out to her. While I was comfortable taking the remedies, I was nervous about taking them in the hospital. I would take them when I would be alone. I wanted to tell the medical staff I was taking all-natural remedies but was not comfortable with doing so. I felt that I was doing something wrong, which I was not. Taking the remedies while I was in the hospital remained clandestine. I felt I needed to protect taking the remedies while also keeping this private. Having to do this did produce some stress. However, I was determined to continue what I had been doing beforehand. When I got connected with my sister on the phone, I proclaimed to her that I was going to write a book so people know there are other options to care for themselves without experiencing any side effects of medication. The homeopathic remedies were helpful to me and were without side effects.

Chapter 8: Ringing the Bell

Finally, being in this room in the regular hospital was moving one step closer to going home. I was now on my own more often, including getting out of bed and walking to the bathroom. I still had to press the call bell for the nurse since it required me to be supervised; however, as long as the nurse was in the room, I could walk to the bathroom. This opportunity led to other activities, such as shaving and brushing my teeth. I was starting to have independence, and the feeling was exhilarating. I could see light at the end of the tunnel, which was my home.

My brother, Michael, was able to visit me, and it was so great to see him. He was relieved to see me as well. With all that happened and not seeing me yet, my mom, sister, and Michael didn't know what to really expect or how I was recuperating from this latest medical drama. Michael was the eyes for Mom and my sister, who lived in separate states from us. He was the updater to both of them after seeing me and spending time with me. My mom especially was deeply comforted that Michael was able to be in phone and physical contact with me and appreciated the extra updates from Michael in addition to Loren's. With all that was going on at this time, it seemed easier to have Loren and Michael visiting me while also not having me overwhelmed by too many visitors. Jack, my son, and Loren were able to see me and spend time with me after the thrombectomy. My children had all been through so much since everything started in March, and now this had happened. I, along with Loren, wished to protect them from having to see and be involved with all this medical stuff. We wanted to limit any feelings of trauma they may have, seeing me, and I wanted to be at my best before seeing them. I deeply aspired to calm their fears and reassure them that I was alright and was going to be healthy again.

I had spent a couple of days in the regular hospital room when we had to make another decision about where I was going to receive Physical and Occupational Therapy. A couple of neurologists came in to check on me and were very pleased with my progress thus far. Both of them mentioned that a plan would be required for in-home physical therapy or outpatient therapy at a rehabilitation center before I could be discharged from the hospital. A rehabilitation doctor came to my room to speak with me and Loren on my options for inpatient therapy or scheduling in-home therapy. My mind was made up; I wanted to be discharged, and I would do outpatient therapy. Loren kindly scheduled my outpatient therapy at Kernan Rehabilitation, where I was for my inpatient rehabilitation therapy after my hospitalization from the previous stroke. Initially, there was talk about me going back to inpatient rehabilitation before going home. Within a short amount of time, I was walking up and down stairs and walking so far with just supervision that I was not a candidate for inpatient rehabilitation and was discharged straight home. When I shared the news with the doctors that outpatient therapy was set up, they discharged me the very next day.

Of course, all of my family members were jubilant with such good progress and that I was homeward bound! And conversely, a bit apprehensive. It was so unbelievably shocking what had happened to me, as well as the speed and ultra heroics to save my life, that all my family members were experiencing various levels of emotional and mental trauma. The discharge from the hospital was also quite quick, considering all that had occurred. This was exceptionally wonderful. However, all of my family was still catching their breath.

In the morning, I was discharged, and Loren called my sister on her way to pick me up. When my sister answered, Loren said she was somewhat nervous about having me home. The discharge from the hospital had moved quickly, and she was feeling traumatized from when she saw me having the stroke, going by ambulance transport to Carroll Hospital, the rapid deterioration of my health, and the time of the essence ambulance transport to get the thrombectomy which pulled me from the door of death. My sister could hear the emotion in her voice and feel the heavy fatigue Loren was experiencing. A lot of compassion arose within my sister for Loren and the space she was in during these moments on the phone. She kept concentrating on sending love, compassion, reassurance and the energy of stability to Loren while they were speaking. My sister was very well aware that

Loren, as the primary caregiver, was deeply in need of support and wished this would be provided in every way possible to her. In her view, from her experiences with herself and her clients, a severe abyss within the medical system is the lack of care, recognition, and healing for the primary caregivers. To her, the medical system is dismal in its awareness that the health and well-being of caregivers are as important as the health and well-being of their patients. Here was Loren, the caregiver, in the same predicament as so many countless others.

On this discharge day, I was super excited to be going home. In all the excitement, we ran into a unique problem: I had no clothes to wear home since arriving at the hospital; all that I wore were hospital gowns. The medical staff found a clean, short-sleeved button-down shirt and track pants for me to wear home. Loren came to pick me up, and a young man came into my room with a wheelchair to take me to Loren as she was waiting for me in the car. I wanted to walk from my room to the car, but the hospital insisted that I be taken out by wheelchair, which I am sure was for liability reasons. Now, here I am on the way home, and thoughts are within me about a new beginning with lots of work ahead of us. I was keenly aware of Loren having to take over all that I was doing around the house, along with having to drive me to therapy and radiation. I had to do the therapy and radiation no matter what. Being home was great, and sleeping in your own bed, nothing beats it. At last, back home.

When I was in the hospital, radiation was suspended. Unbeknownst to me, my mom, sister, and Michael were relieved. As much as others were saying the radiation was not a cause of the two strokes and the blood clot, they did not believe this. All that had just happened occurred during radiation, so believing that radiation was somehow not implicated in this dramatic medical episode that almost claimed my life was implausible to them. They were palpably feeling that they nearly lost me as their very much loved family member and were still feeling that I was extremely vulnerable and that something else could happen to me. They also had unique experiences along with me around my father's last couple of worldly years in which the renegade cells of Multiple Myeloma were found within him.

As I have shared before, my life and my father's life have shared strongly eerie parallels. My father was having chemo and radiation for this renegade cell situation. Revlimed, the "gold standard" chemo, other than causing 100-pound weight loss after six weeks of it, heavy mental depression with anxiety, sagging skin from the weight loss along with his skin peeling that my mother had to wash off of him in the shower, had failed. The renegade cells returned quickly and more ferociously. Now, it was Velcade, another chemo along with radiation. The combination of Velcade with radiation was impairing my father's intestinal function, and no one in the medical community was watching or assessing. I had flown to see him a week earlier for three days, and Michael came after me for three days. My sister gave us space to connect with our father, heading home for three weeks to tend to personal matters and allowing us cherished time with Dad.

On the night she was taking a red-eye flight back to stay with my father and mother for three weeks, her cell phone rang as she was boarding the plane at 11:30 pm. It was mom, "the ambulance is on the way. Dad's abdomen is as big as a football. He hasn't urinated all day. I have to go." This was a complete time-of-the-essence emergency situation, and the medical team had to act fast. As my mom hung up, instant nausea and grief washed over my sister like a tidal wave as she was showing her boarding pass to the flight staff. In disbelief and feeling like she was floating as she was getting her seat, she knew my father had a complete intestinal and urinary blockage that could be deadly and kill him very quickly if his intestines and/or bladder burst. By the time she arrived, he may be no longer alive. By fortune, no one was sitting next to her. She lay on the row of seats for the entire 3 ½ hour flight with a mala in her hand and mentally chanting mantras for his protection, long life, health, alleviation of pain, and that he would not become septic. She also focused on stabilizing her emotions and thoughts so that she would be able to be of the most benefit when she arrived at the hospital.

Like me, with this past medically urgent situation, the emergency room staff was waiting for him when he arrived at the ER. The medical team worked fervently to evacuate 3 litres of urine that had accumulated in his bladder. It was astonishing that his bladder somehow held this immense amount of urine and did not burst. However, the bladder was massively spasming after the release of the urine, and he was in intense pain. Next, they were helping my father get the intestines going and release the complete intestinal blockage from constipation quickly. Fortunately, this was accomplished before any intestinal rupture. However, he was very sick, in acute kidney failure with dark brown urine from the previous urinary blockage, and on the edge of sepsis. The Velcade, along with radiation, had helped his intestines slow down as well as develop severe constipation. Now, his life was at risk. His health wavered between becoming septic and then being able to shift course and not be septic. My sister literally lived day and night with him in the hospital, just taking a break to go home, shower, sleep in a bed for 1-2 hours, eat something, and return. She saw the exhaustion, sadness, and worry within Mom and needed to relieve her as much as possible to help care for Dad. I went to visit him in the hospital and Michael at a separate time from me. It deeply rattled me to see my once very strong 240-pound, stocky-built father in such a condition. He spent 10 days in the hospital recovering and 7 days in rehabilitation, which included still being hospitalized till over Christmas because he had lost so much strength and was depleted from all that had happened.

From this harrowing experience, my mother, Michael, and my sister were now living in my situation. In their minds, radiation, two strokes, a severe brain blood clot, and my health overall weakened. A repetition similar to my father along with the feelings that something more could happen to me. When the blood clot occurred, initially, both Michael and my sister were adamantly against any more radiation. They felt harm was happening. The renegade cell mass had been shrinking before any radiation and now was already shrinking even more rapidly from what was seen on the last MRI at Carroll Hospital. They wished for more recovery time for me and my brain from all the insults and injury. They also felt more time should be given to see if there would be more shrinking and resolution without radiation. Each of them had texted me and asked me to consider pausing radiation, and Michael directly shared this with me when he saw me in the hospital. My mother had been deeply contemplating almost losing me in her life, a son she loved dearly, and also started having the feeling to stop radiation and texted me her thoughts about this.

When I was discharged home, I was on heavy blood thinners to prevent the possibility of more blood clots developing. Besides taking blood thinner medication daily by mouth, I had to inject another blood thinner medication into my abdomen every day. My entire right forearm was a deeply bruised black color from all the IVs while being on these blood thinners, and my abdomen had multiple small bruises from the injections. The black color bruising on my arm healed very slowly and got better more quickly after using some homeopathic remedies. All of this was a lot for me to contend with taking these new medications along with all the others that I took on a regular schedule for my kidneys. Although I was so happy to be home, I was very debilitated as well as overcome by everything that had just transpired.

The day after arriving home, I resumed radiation again. The time frame to complete the radiation was extended due to me being in the hospital. Admittedly, the schedule felt daunting. I just wanted to ring that bell, finish this stuff, and get done with these leftover sessions. There was the "magic number" of 25 sessions. Loren and my children supported this, but Michael and my sister did not, and my Mom was very apprehensive about continuing at this time. They were also experiencing "negative signs" that I was very vulnerable to another stroke and very possibly would not live if this happened. They did not share with me at the time what was going on with them, for they did not wish to create worry in me, or I would know they were not supporting this current plan. Instead, each of them focused earnestly on prayers that an intervention would emerge and somehow I would not receive radiation when I was not supposed to because it could create a devastating effect. I had no idea they were doing this and only found out long after radiation was finished.

When the blood clot in the brain occurred, my sister felt such darkness around her, and when she "looked" at me in her mind's eye, I looked dark greyish, and she couldn't see or feel my future. She would wake up in the middle of the night with tension in her heart that something was so very wrong and could not shake the feeling. The day after the thrombectomy, she called my mother to check in on her. They had been talking daily for months as my sister was really concerned for her regarding me. They were talking about this medical regime I was partaking in, the strokes, the blood clot, and then the plan to resume radiation right away. Exasperated, she emotionally exclaimed to my mother, "My father was not a statistic! My brother is not a statistic! They are precious, beautiful, unique human beings! This world of insanity within this medical community helped bring so much additional suffering to my father besides that disease, and now the same happens to my brother!? I have to watch this all over again!? One size does not fit all! He doesn't fit their so-called protocols, and they are playing with his life because he has to fit into their limited minds, which can't analyze any other possibility! Constantly, these doctors thread and tread on life and death with their so-called treatments!" All mom could do was listen. She knew my sister's heart was over wrought like hers. She replied, "I am very concerned about the radiation also. I don't think it's good for him at this time. All we can do is pray that he is protected, that the right things will occur, and that he will be okay. We must trust that our prayers are helping and will help take care of him." My sister agreed and knew all of this was true.

With me starting radiation so quickly after coming home, a reoccurring memory repeatedly pervaded my sister's mind about the day after my father returned home from his difficult, prolonged hospitalization. While in a different room, she overheard my mother say after answering the phone, "No, we are not interested in any trials."

Instead of the oncology practice asking how my father was doing and checking in on him being home from the hospital, they were pushing some other intervention. A few minutes later, the radiology center that had missed the signs of a probable intestinal blockage called and asked my mother if my father wanted to continue radiation, to which she replied, "No." After all the medical trauma that my father had been through, that my mother and my sister had been a part of, now only being home 1 day, these 2 phone calls. My sister felt simultaneously sickened, shocked, and appalled while asking herself, what has the medical system become, and who were these people? The parallel of what happened right after my father got home from the hospital and now me starting radiation again the day after getting home was very disconcerting.

She was filled with sadness for me and also especially my mother that something else could happen to me and that no longer would they have me in their lives. She also couldn't imagine this happening to Loren and our children. She has a way of doing what is known as divination to look into a situation, and they are not showing good results. She also consulted with spiritual people she knew who were seeing the same thing. On a phone call with my mother, when they started talking, the sky where she lived suddenly turned dark. While they were talking about my situation, particularly about radiation, a sudden and intense storm erupted outside, marked by a brilliant flash of lightning, a loud clap of thunder, and a torrential downpour that seemed to come out of nowhere, dramatizing the tension in the air. She knew this was a foreboding of a possible future and felt so heavy and depressed. My mother was grief-stricken, saying that she did not know if she would live on if something happened to me. All signs that all was not well.

Michael was not faring any better. His thoughts were consumed with another stroke that could be imminent, and maybe this would end the connection between us because I would leave this world. Right after the two strokes happened and I was in Carroll hospital, he called my sister. He was feeling freaked out from what had just happened while he was alone in the apartment, sitting on the couch. It was completely quiet during the middle of the day, and he was doing some work on the computer. Without warning, a large picture above him suddenly and mysteriously unhinged from the left side of the frame, swinging violently back and forth like a pendulum while hanging by the picture hook on the right side. He had personally hung this months ago and ensured it was completely secure on the wall. He felt like something had actually lifted the picture off the wall with a force that created the rapid swinging back and forth. The large picture was swinging so hard and wildly that it made a loud scraping noise against the wall and left a deep mark on the wall. Immediately, my face and the thought of me came vividly into his mind, and with it, a thought struck his mind, "something serious is going to happen to Matt."

When Michael relayed all of this to her, she became full of dread. My sister thought it was an omen about me and that something else was going to happen. She was very much hoping this was a wrong conclusion. She wanted to believe that the worst was finished after the two strokes occurred, although a picture becoming loose in such a way from the wall definitely appeared as a warning. The next day in Carroll Hospital, after the two strokes, a blood clot in the brain emerged. When Michael visited me in the hospital after the thrombectomy resolved the blood clot, he was so happy to see me, and when he heard me say, "Let's get going with the radiation and ring the bell," his heart sank. When he was in the car after leaving, he called my sister and said, "It was just within minutes of his life not being here anymore, and he was talking about ringing that bell. He is still in the hospital, just out of ICU and talking about that bell. I can't believe it."

When I resumed radiation after returning home, something else had transpired with him. Michael and his girlfriend, Laurel, regularly light candles in their apartment. They had a large candle on the coffee table in front of the couch. He lit the candle with a match, blew the match out, and put it aside on a glass coaster where he always put the blown-out matches. He had been sitting on the couch in front of the table when Laurel appeared and said, "Please don't leave a burnt match on paper like this. It's not safe." Michael said, "What paper and match?" Laurel pointed to the table in front of him and said, "This paper and match." Shocked, he responded, "I didn't put that match or paper there." Somehow, it had appeared on its own with him sitting there, and he never noticed until she said something. The image of fire came into his mind, and then radiation, like being a source of heat. "It's about Matt," he thought. Between the paper and the burnt match on it, along with this thought, he had chills all over his body. He immediately texted my sister, "Something really disturbing just happened. I need to call you." he relayed this to her when they talked. She said, "we are playing with fire. We need to do much more mantra and prayers. His life is in a precarious balance."

They did not disclose any of what they were experiencing to me and just kept adamantly doing their own prayers and mantras for my safety, protection, and long, happy life and that something would intervene and stop the radiation sessions because they felt so deeply my life was being threatened. There had been a few radiation sessions in a row already, and with the next week that was coming up in particular, each of them had feelings that my body was too weak to endure any more radiation. On their end, things felt desperate. Then, spontaneously, the divine intervention arose. The radiation machine broke down. My sister even remembers it was on a Wednesday. From their point of view, the divine solution had appeared. My mother, Michael and my sister felt gigantic relief that there would be no radiation on Wednesday, Thursday, Friday and then obviously, over the weekend. Five days off. Although they thought my body was very weak, they felt like the time off gave me an incredibly necessary and vital respite, and now I would be able to endure the sessions until the number 25 was attained.

On my end, not knowing about all this was going on, my goal was to finish radiation in a couple of weeks to ring the bell after my final session. The proton center had a large bell that each person rang at the conclusion of the radiation sessions. My mind focused on ringing that bell, being finished and putting all of this behind me to move on to my future. During this time, for the most part, my days were either going to radiation sessions or to outpatient therapy appointments.

Loren was working while also driving me to sessions and taking care of our house and other tasks. Lucy was in high school and dancing classes. Jack and Georgia were working. Molly and Brad were living in a separate state. Everyone was trying to balance their lives all the while as therapy and radiation sessions were going on with me. This was such a challenging time for everyone. Lucy was in her last dance competition before college, and Loren was going with her to a 5-day event. Loren was troubled about leaving me, and so was Lucy, and I really wished to be able to support Lucy at her dance competition. Loren wanted to keep Lucy's life as normal as possible, and going with her to her last dance competition before college was an important priority. Our daily life was being deeply interfered with by these personal health dilemmas plaguing me, as well as the time required for therapy and radiation.

I was very fatigued, and going back to Kernan Rehabilitation outpatient clinic was depressing for me. I felt that I was moving backwards, having to be back there. I was hard on myself when I didn't score well on a Speech or Occupational Therapy task. My mind wanted to prove I didn't need the therapies. It was difficult to accept this most recent setback. Loren would talk to me after the therapy sessions to remind me I'm not perfect and that this is what therapy is for. Deep down, I knew she was right, but my stubbornness and strong mind would not allow me to forgive myself for my mistakes. I wanted to be done with therapy.

I had been moving towards the final destination, the last radiation session, which seemed so far away but had finally arrived today. Getting up from the radiation table for the last time felt like, "ok, we have hit this goal!" and it was like music to my ears when I heard it was time to ring the bell. I sat on the table for a few minutes to let the dizziness dissipate and let it soak in that I did it. It's Finished. It's Done. I've Completed. A nurse went to the waiting area to bring Loren into the room. When Loren arrived, she walked up to me where I was sitting. With a big smile on her face, she took my hand and helped me off the table. What a huge accomplishment! I held Loren's hand tightly as we walked down the long hallway because I was very dizzy from lying flat on my back during the session. I was utterly ready to put the renegade cell mass entirely behind me, totally vanquished. Totally vanished.

Finally, the day came, session number 25, the final one. Loren and I walked together to the big bell on the wall. A medical staff member was with us and took a short video of me ringing the bell. I took the long cloth strap attached to the bell and rang it seven times with my left hand. Overjoyed, Loren grabbed my right hand, and we lifted our arms together into the air. She was holding my "certificate of completion" in her other hand. The medical staff filmed ringing the bell, and Loren cheered. After finishing ringing the bell, I placed my right arm around her waist and we kissed each other. This is now over. Totally concluded.

We left the proton center and drove a couple of blocks to the University Medical Center to meet with the renegade cell doctor to see if the radiation did its job of eradicating the mass in my brain. I was feeling nervous about this appointment, but I wanted to know how things were after 25 radiation sessions. The only thing I wanted to hear during this appointment was the mass was gone. Loren reminded me, she said, "Matt, the mass is not going to be totally gone. They said it there would be a scar." I had been sincerely and profoundly praying for this to be absolutely no more, and this is what exactly happened. The renegade cell doctor reviewing the MRI said, "This is remarkable" because the mass had disappeared from the screen he was reviewing. He kept looking at the screen on the MRI to see where the mass had been since he could not find the mass where he expected it to be. He was pretty sure it was in this one place, but he could not find evidence of it. Obviously, I was tremendously excited by this news! I was aware that I was under an umbrella of protection from my Lord Almighty, along with all of the assistance I had been receiving to get to this point. The homeopathic remedies, beneficial thoughts and prayers from many people, healing meditation sessions, sessions to help healing energy within me express more easily and readily, love from family, friends, and colleagues, therapy sessions, and the outstanding care from my medical team.

When we got home, there was that last thing to do with the "radiation paper chain" hanging on a door that my daughter Lucy, had lovingly made for me that also helped keep me and my family motivated. When I walked into the house, I walked to that last paper link, raised both of my arms and undid that link. We were videoing this moment, and I turned around with that torn paper link in my right hand, waving both arms in the air. I was really fatigued, and it was too much effort to speak. As I waved my arms, Loren said, "Yeah, last link on the chain. Way to go, Babe." Georgia was there, too, cheering for me.

Throughout all this time, by the blessings of God, I had my sister, Michael, and my mother constantly reminding me that healing from what I have been through is a process, taking one day at a time, and I was making progress every week. Though they did not see me every day, we did speak on the phone practically daily. My sister and I would have Zoom and FaceTime calls to do Occupational Therapy exercises and meditation. She also worked with me "energetically" and helped me access the healing and health energy within me. I couldn't necessarily see or feel this especially due to this decreased state of health I was going through. She kept the knowing and vision of this inner abundance within me and helped me facilitate its presence and increase. This is important because oftentimes when one is doing the right behaviors, expending energy with good motivation, and not seeing the results they are hoping for, many times, people give up.

My mother, Michael, and my sister reminded me how far I had come and that my voice was stronger each time we spoke. This was refreshing for me to hear because I felt I was working hard and not achieving the results I was hoping for. They consistently gave me the reminder that after having as many strokes as I had, as well as going through radiation to eradicate the mass in my head, I was on the right track, and the healing process was going to take time. Mom designated the four of us as Team Blake, and we were all in this together. They would be the cheerleaders for even the smallest accomplishments, and I was to accept these for what they were. It was another step in the process of healing and restoration of health.

After the completion of radiation, I continued with therapy and recovered from this roller coaster ride of a summer. On the 5th day after finishing radiation, my mother came for a visit to see me, see my family, and help out in any way during her visit. She was so relieved and happy to put her eyes on me. We spent a lot of time talking, and she was helping me to let go of mistakes, of feeling guilty about everything that had happened, and to forgive myself for not being able to support my family, which was weighing very heavily on my mind. I was very exhausted, had little appetite for food or desire to eat, and had lost a lot of weight. I did not fully realize that I was also being affected by radiation side effects besides the strokes and the blood clot that happened in the brain. It was very reassuring that my mother was here, and she also spent time with Loren as well.

A month after the radiation was done, it was time for the MRI to check the status of the renegade cell mass as a follow-up. Fervently I was praying the result would be very good. Most definitely, all my family was praying for the same thing. When the MRI report came, the results were actually amazing. Loren texted the entire family, *"Good news from radiology. The tumor is negligible!!!! The doctor said the results were dramatic. The full report won't be in for a while, but the doctor is very optimistic!!"* Later in the day, when the MRI was re-looked at by the renegade cell doctor, Loren texted me and said, *"tumor is totally gone!"* The doctors were very shocked because the expectation was that the renegade cell mass would leave a scar and not be totally gone. I felt extremely happy that this renegade cell mass was extinct. The doctor said this was very unusual and in a good way. Everyone in my family was ecstatic. My sister shared with me by text that day that she really expected the renegade cell mass to be completely gone. She said the homeopathic remedies, specific meditations, and our working together to help access additional healing energy within me were altogether very large contributors to this happening. Okay, a very positive MRI and health result. Truly amazing. Now, onwards, I thought, as Loren and I drove home after this entirely eventful day.

We usually think that obstacles are present in our lives as ways to divert us, distract us, confuse us, frustrate us or defeat us at worst. Obstacles feel very difficult, impenetrable and often stronger than us. They push back against us, try our patience, create misery, produce fatigue and usually make us feel weaker than "they," "the obstacles." Obstacles deplete our energy and, therefore, we experience a lack, a feeling of less and possibly loss. This lack appears as some sort of poverty or lack of abundance in our lives, be it money, material needs, health, personal relationships, our sense of purpose and being or our professional work. This is the typical way of thinking about obstacles that most of us humans possess. After all, this experience with obstacles has been repeated over and over with people throughout generation after generation. It appears engrained within our psyche as being human. Why would obstacles be thought of in any other way than this? Based on my experience, especially with a kidney transplant, strokes, blood clots, and renegade cells, I would have to say that this is "wrong behavior" that will undoubtedly foster decline, misfortune, or worse. How in the world can I say this? How can I make this claim?

I say this because, as can be seen throughout this book, obstacle after obstacle popped up sometimes without warning and with possibly delirious detrimental effects and results. Obviously, I have lived through these obstacles and have sustained continual progress forward. How? Even though I wasn't always consciously aware of doing this, my mind was looking at a way to go around, go over, go under, go through, step on, crush or dissolve the obstacle facing me, or even directly challenging me at the moment. By doing this, my mind was assuming a path that was certainly stronger than the obstacle presenting itself. What is this? This abundance of "spiritedness or vitality or potency" within and around us is ever-present. It appears hidden, although this is an intricate and intimate connection that exists for each person regardless of country, background, culture or identification of ourselves. The "abundance" is not limited or depleted and has a consistent and constant flow. It is subtle, however, very compelling, and sometimes it will manifest or be displayed in an obvious way, such as what we may see as a miracle. People will experience this abundance however their mind "sees, feels, or perceives."

This is the ultimate right behavior that I speak about. When we are confronted with a challenge or a crisis, if our minds are focused correctly, we will employ and engage in the right behaviors like a proper positive attitude, creating a healthy diet, diminishing or stopping negative emotions and thoughts or becoming a better person overall. The conglomeration of these right behaviors gels into a powerful energy that can be channeled and then directed towards accomplishing. And, very frequently, when this is done, a solution to the obstacle will come forward, or the obstacle will disappear. Other times, we are utilizing the right behaviors, feeling we are doing all we can and either not getting anywhere or getting very little improvement or positive results. It's entirely beneficial to recognize that the ultimate right behavior, the abundance of knowing, this mindset, is what we contemplate, draw on and allow. In actuality, this subtle, seemly hidden abundance is the never-ending spring that sustains us, nourishes us, provides for us, and cares for us.

At the times in which the obstacle seemed so powerful, unyielding, and unmoving on this very rough road with my health, such as the stroke in March, the lumbar puncture, the other two strokes, radiation, and the blood clot, my mind deeply delved into this abundance, and stayed there. Holding onto this provided the fortitude, endurance, patience, confidence, and security to be within this obstacle while also the path around it, through it, under it, over it, stepping on it, or dissolving it revealed and was achieved.

We humans are all interconnected to each other. Although we present in many different ways depending upon different conditions, such as where our family origins are from in this world, we are interconnected. We all have DNA and the human body's physical systems and organs. We rely on each other for sustenance in a variety of ways and social connections. As humans, we are also interconnected to nature as we are unable to live on this bountiful planet without nature. When facing an immense obstacle, even one that can take our life, interconnection is a potently forceful antidote. We are physically interconnected through having the same body systems and body parts. Also, our mind energy is subtle, and we can interconnect through our minds. When we think of someone, that person's mind does have some sort of connection with our thoughts. For this reason, positive affirmations, prayers and other positive actions powerfully influence a situation, and this is why they work.

When a dire obstacle was happening to me, like the stroke in March, essentially, this obstacle became weaker and weaker over time while other very constructive and helpful circumstances or opportunities came into its place. As a result, I had demonstrable and tangible improvements. So many people were praying for me and thinking positive thoughts for me, even if they did not know me. As I expressed before, I felt this in the hospital, and it was very reassuring as it propelled me into deeper healing. When the praying and thinking positive thoughts were happening for me, I knew it was also bringing the best of other things needed, such as the correct and best doctors, the kind and knowledgeable nurses, the right hospital, a person taking care of my business and love holding my family members together. It's the abundance of interconnectedness that helped me so much during that hospitalization. That obstacle of that stroke could have been so much more severe and longer lasting with very slow progress. Instead, this obstacle was dramatically weakened and cut down while abundance prospered and beneficially assisted me.

I saw and experienced these right behaviors and an abundant mindset, along with interconnection, all working to defeat the illusion of lack or taking away from my health, and I unquestionably reaped benefits. Time and time again, I was profoundly grateful. Now, finally, with the radiation completed, the next chapter of my life was unfolding. It was time to reassume further being a husband, a father, a brother, a son, a friend, a business person, and most importantly in order for me to do all of this, a loving and trusted friend to myself.

Chapter 9: Propelling Through the Mirages of Obstacles

I had propelled through the 25 radiation sessions. I was going home now to be with my family and dogs. I had the thought, "It's time to go home, now what?" My mind contemplated, "What is my life going to become?" At this time, for sure, the simplest tasks were now complex for me to do. Nothing was easy. I had lost a lot of weight because I had no appetite. I would literally cringe when it was time to eat. The only thing I really liked were chocolate peanut butter protein bars. I could only eat one bite at a time very slowly. I was hoping that these protein bars would give me enough protein and calories to gain weight, but in truth, it was very unrealistic. Although I was extremely tired, I could not sleep well and constantly felt unrefreshed even if I had been able to fall asleep. All of this resulted in me staying in bed for 12 to 15 hours a day, which, besides still feeling very lethargic and worn out, left me feeling guilty for being in bed for so long. Thoughts would occupy my mind, such as, "Where do I go from here? How can I get going again?"

Everything was so hard, and I felt like no one really understood what I was going through, which left me feeling very alone and not understood or misunderstood. It was also not easy to communicate how I was really feeling. I did not fully comprehend how I was feeling; everything that had happened felt confusing, and I was overcome by it all. I was still in recovery mode from everything that had happened to my brain, my body and my mind. Putting thoughts together was not easy, and conveying them was even more difficult. Processing in my brain still felt disorganized and slow. There would often be a gap or delay between phrases or sentences I was communicating and speaking, which was due to the previous stroke in March. The gap or delay had improved; however, it was still there. The neurologist in March, when the stroke occurred, said to me that there was expected full recovery within 6-9 months. This declaration was before the two other strokes, a blood clot in the brain, and 25 radiation sessions. People around me did not seem to understand that we were no longer on this 6-9 month full recovery timeline. A lot of unexpected events had occurred, which had now deviated from this prior trajectory. My sister, being a previous Occupational Therapist with a neurological background, did get it and consistently reminded me of the extraordinary progress, recovery, and restoration that was taking place.

Not sleeping was affecting my everyday life. It's a different tired with someone who has had a stroke in March, two other strokes, a brain blood clot, 2 hospitalizations including a rehabilitation stay within four months, and just finishing 25 radiation sessions. At this point in time, I had lost over 50 lbs since the first stroke in March, and time had blended together because now it's mid-July. I struggle to get through each day since I am not thinking clearly, and my body and mind are wholly exhausted. There were very dark moments of thinking that it probably would be easier on everyone else if I just died. What kept me going was the feeling that there must be some motive for why I am still here. When I looked more deeply within myself about this, there was a knowing that God has a plan for me. I sensed that my life was saved to get to this point. Keeping focus on this and reminding myself was essential to guide me through this heavily arduous time.

I was still on a homeopathic remedy schedule, and I confided in my sister I was not sleeping well and, as a result, was unfocused and constantly feeling weary. We decided on some remedies to help me sleep more easily and stay asleep during the night rather than waking up with my mind ruminating for hours. For a number of nights, we made a plan that I would call her when I was in bed, getting ready for sleep. Together, we would do a nightly guided meditation that she would do to help my body feel relaxed and then for my mind to slow down thinking, which would help me to feel calm so it was easier to go to sleep. The remedies, along with this, helped me tremendously to be able to sleep more deeply, and I started to feel better in the morning. Over time, I was able to be in bed less time and get up earlier. I also had a routine in the evening to listen to guided meditations around Jesus and the 2 of us being together as I continued traveling through everything since last March when everything started. My mind and heart felt comforted we were walking this together.

I was trying so hard to get better as quickly as possible. The needle on the meter of progress seemed at times to be at a snail's pace forward and sometimes seemed to go back to where it was before. Often, this was met with frustration coupled with a feeling of being let down. During one of our conversations, I shared with my sister that I was trying so hard to be able to do more things better and more efficiently. She responded to me, "Why are you trying? Trying implies still in a struggle, still not attaining and not getting to success. It implies a lack and can't be part of the abundance mindset. So just do. Always think, 'I am doing it,' whatever it is, or 'I am doing this.' Then you are in the action of doing, and when you do, there is movement, there is change, and there can much more easily be a success."

I thought about this and said back to her, "Yes indeed. I am doing it!" This became my slogan, and I repeated it daily when doing any task, which made things easier to accomplish.

When doubt crept into my thoughts during a conversation with us together, like, "Do you think my memory will get a lot better?"

She retorted, "Why not?"

Hearing this would make me stop my current train of thought and think about "why not" instead. "Why not?" created more potential thinking like, "Well, maybe this could be possible. Why have I automatically assumed it is not possible?" When my thinking shifted to this, I felt more courage and a renewed, energized feeling to keep going and then to reaffirm, "Yes, I am doing this." Once again, my mind was much more within the abundance mindset, which could only benefit me.

Team Blake of mom, me, Michael, and my sister had and still do have a group text. Team Blake knew that "I am doing this or I am doing it" was my repetition to increase courage, clarity, and determination to go over, go around, go under, or step on what seemed to be holding me back or in my way to succeeding. Every Friday, we had the Friday Team Blake Huddle (FTBH). From being a therapist, my sister knew that when she was able to use "language" that had definite meaning and feeling to who she was working with, therapy became much more triumphant for her client. She knows mostly nothing about football and has no interest in it. However, she knows our family most assuredly does.

FTBH was the weekly check-in. She, as a team "cheerleader," would text from our group text for me to send the report of success updates, no matter how small. *"Hey there, Team Blake! It's time for our weekly FBTH (Friday Team Blake Huddle) this week! Dearest precious Matt, we are looking forward to the accomplishment updates, no matter how small." So, let's see what the checklist has this week!"* I would text the success update "stats," and Team Blake positively recognized them. Mom, as our founder and "captain" of Team Blake, always texted something inspiringly back such as, *"Great Report Matt! Very proud of your progress!"* Michael, the team "coach and trainer," always sent out the thumbs up emoji with positive energy supporting our team. Our Team Blake check-ins really sustained me, and I looked forward to regularly sharing the "weekly progress" reports.

From my business background, I was cognizant that for me to feel more confident that I was getting better, I needed to see and be reminded of the positive changes. I established a journal of little successes, no matter how small the success. I wrote each day what I had been able to do, even if it was a repeated success from the day before. The journal was a reassurance that good changes were happening daily. It helped keep me going even if a day wasn't the best and I was feeling down. Living in such an advanced technological society as ours does have the flaw of under appreciating the simplest tasks of our day. For me, I intensely appreciated waking up and getting out of bed, arranging my clothes and getting dressed, and being able to remember my day before and what I would be doing today. For anyone facing an obstacle, such as a health problem that is negatively impacting their life, it is essential to track and acknowledge the small successes. Carrying out this provides you with a perspective of doing, changing, advancing, and attaining that cumulates in propelling ahead, leaving the "previous" behind, and moving beyond.

Having my own business became too much for me to handle; it was overwhelming and, unfortunately, impossible. Like I had conversed with Loren and my daughter, Georgia, after I had returned home from the rehabilitation center in April, once again, we talked about my business. They convinced me that if I wanted to continue being a sales trainer, I should look to merge my business with another franchise. At this time, I felt like I was giving up on my dream. However, one thing I learned previously was that to be an effective leader or salesperson, it is essential to make decisions that are best for my family and my future. As well as not making any decision was also, in effect, a decision that kept me stuck and not able to move forward. It was an enormous decision to shift into a merger because I believed that I was going to heal fully, as it was only a matter of time. I thought many of my clients would be happy to wait for me to get better; however, I also know that they have businesses to run. Since I could not do any training, I looked at how to help them to receive training during this time. Fortunately, another franchise took in my clients to provide training while I was healing and regaining and restoring health.

Initially, all of this hurt my pride, and I felt that I was losing everything and that there was nothing I could do. The obstacle of losing my business, my clients, my purpose to help others through training, and my identity to all of this was looming large over me. Discussing with both Loren and Georgia gave me the resolve to contact other franchises to see if we could merge our businesses together. This outcome would allow me to continue to do what I love while also having the support necessary for me to recover, regain, and fully return to what I enjoy greatly when possible. I have learned that leaders make decisions, especially when it's hard to make decisions. My mind persisted with complete confidence that God was with me and that this business decision must be a part of a plan. Somehow, everything would connect together even though things did not go as smoothly as I expected.

The other franchise welcomed me with open arms, which I exceedingly appreciated. However, many of my clients had moved on, and in lots of ways, I was starting all over, which was now very daunting. I was at the beginning gate bringing on new clients. Now, I no longer have the office that I gave up a few months ago; I am working from my home office. There was new technology knowledge to acquire and accomplish. Since I was not driving, all meetings with team members and clients were now remote and virtual, which I was not used to. Communication with team members was also through the computer on technology platforms that I was not familiar with. Planning, organization, coordination, structuring, and strategizing were necessary requirements in this business, which was not easy to enact at this time. I was striving to get up to speed with the new training programs that were being delivered through Zoom to much larger training classes than I had ever done before.

Even though I would have preferred working from the business office I had before, I did not have any desire to go outside the house. It was difficult enough to get out of bed, shower, get my clothes on, and think about eating something for breakfast, especially without an appetite. It felt like too much was going on with just being at home. I actively kept thinking in my mind about what I was doing at the moment and how to do it, even for the most basic and simple things like how to get my pants on correctly. When a task was finished, I had to think about what I was going to do next and how to do the next thing. The idea of leaving our home to even go for a walk or go out for coffee with a friend or family was overwhelming. If I would run into someone who knew me when I was out somewhere, it was difficult and uncomfortable to answer the questions they were asking. I was in uncharted territory with everything that had happened and was now going on. I felt uncertain about myself, how I was doing, and how things would be in the future. I really did not know how the person who knew me before was seeing me now, which was also a unsettling feeling. As can be imagined, functioning like this was overly consuming and exceedingly draining.

The times that I did get out of the house were for therapy appointments or medical appointment follow-ups, with Loren driving me around. This was my life "out of the house." Not being able to drive made me feel helpless, dependent, uncomfortable, and feeling down that I was not able to drive myself. In therapy, at times, the task being planned for me felt questionable if I could rise to the occasion of what was being asked. We were doing a variety of yoga poses to increase strength, coordination, endurance, and confidence in using and moving my body. I would say to them, "I am not sure even a healthy person could do this, and I am definitely unsure that I have the ability." They would agree that this yoga pose may be difficult for a healthy person; however, they believed that I had the capacity. With each trial of getting into a yoga pose position, I found I was able to be successful.

My brother Michael had been trying unsuccessfully to reach me more consistently on the phone. With my schedule filled with appointments and my energy drained from daily struggles, plus prolonged periods of rest, connecting with each other became impossible. Michael wished to help me and was frustrated we could not talk or meet more easily. He did not give up and just kept trying. One day, Loren called him and asked him to come over and talk with her and me about getting stronger beyond therapy. Michael is a personal trainer, so he was perfect for helping me attain more strength, coordination, ability, and self-assurance.

We decided that Michael would come over to our house a couple of times a week and start doing personal training sessions with me. I absolutely wasn't ready for the gym. However, this was the future goal. When I first got home from the rehabilitation hospital after the stroke in March, he stopped by the house to see me. I was sitting outside with him, and he said to me, "Hey, look, when you're ready, you and I are going to go the gym together. You don't even have to do anything the first time we go; you can just be there, get used to being there, meet people, and get comfortable. We will just do simple things when you are ready. It will be very easy, and you will be able to do this." At the time, it felt really good to get back into a gym again felt thrilling and made my spirits soar. I had not been in a gym in such a long time. Now, in my present state, the ability to do this is not anywhere close beside the fact that I do not even have the ease to leave the house, much less go to a gym.

After this meeting at the house with Loren, me, and Michael, the time came to begin some sort of formal home workouts with him. In the beginning, nothing was easy-peasy with these home workouts that Michael was creating and doing with me. It took all my energy, even if I had to do the simplest exercises. I fatigued very easily, so I required quite long rest breaks. I moved slowly, deliberately, and methodically, unsure of my body's movements, and I easily got dizzy. Obviously, my body was in a very unconditioned state. Everything took great effort and energy to do. I remember the day I first did a modified push-up on the floor. It took a lot of strength, and I got very tired fast, but I did it! Michael sent a video of this to my sister and mom. Both were joyous with this accomplishment, and my sister, as a therapist, knew from where all this started with me originally, and she had tears of joy in her eyes while viewing the video. While watching the video she remarked to herself, "How far he has come in a short amount of time from everything is amazing! I am so happy for him! This is so wonderful that Michael can train with him physically!"

My first "formal outing" from the house also was with Michael. There was an opportunity to meet with someone who could possibly become a client in the future, whom I was going to meet at a coffee shop, and I had arranged a ride. Michael and I went out together to "practice" getting in and out of a car that was different from Loren's, getting into the coffee shop and navigating the space, being able to order and be in the coffee shop with people, furniture, and other noise around. In my current condition, I found this entire experience more difficult than I thought it would be, which was discouraging, to say the least. I was not as proficient as I expected. Rather than boosting my morale about myself and feeling like I could move forward being out of the house more, I found that I was more comfortable and assured of being at home instead. As things always seem to work out for the best, the meeting was postponed.

After a couple of months, a chance business opportunity presented itself. A business networking conference was being held in Orlando, Florida, for three days. Over 2,000 people would be there, and I would be flying by myself there and back. This would require me to navigate through 2 airports, get onto a plane, and successfully get to the hotel where the conference was being held. I gave myself a pep talk, "I can do this. This is an opportunity that I should not miss." I considered pushing myself a bit to maintain the momentum of my business with the partner franchise. Although attending this conference was enticing, I was unsure about going.

On the day when I arrived at the airport to fly to Orlando, I was managing to get through security with my carry-on bag. A thought struck my mind, "In my current physical condition, I am not able to defend myself. If someone came up to me, they could knock me right over." Although this was not going to come about, there was an undeniable and uncomfortable awareness of this susceptibility. When I got onto the plane, I could not lift the carry-on bag into the overhead bin due to my left arm being weaker. To do this, I was given help, something without a doubt I was not used to.

Happily, I arrived in Orlando and went to the hotel. Every step of traveling to and being at this conference was literally one step at a time, being patient with myself and trusting that I could succeed. It was an enormous hotel, and the elevators were quite confusing. I had to know which elevator to get into in order to get to where I was going. Again, the repeated thinking with whatever came next during being here, "Can I do this?" And then, the repeated answer, "I am doing this." To my delight, I had the ability to attend each day at the conference along with nightly events while negotiating to be with such large crowds of attendees. It was great! After returning home, I was exhausted from the overstimulation of traveling, being away from home, and attending the conference. At the same time, I was really pleased with how well everything went.

Looking at my life all of this time after the radiation sessions, it was like obstacle after obstacle appearing, one after the other, sort of relentlessly. Constantly feeling tired with lack of energy, having to literally think how to wash my hair in the shower and other essential daily activities, forcing myself to eat due to minimal appetite, seemingly endless therapy and medical appointments, reconfiguring all my business, difficulty being able to communicate with and spend good time with my family, lack of financial wealth, experiencing physical debility, emotional ups and downs and so on.

While these were transpiring within me and around me, I had a reoccurring, consistent motivation. As I reflected on my recent travels, watching myself navigate, handle, manage, take charge of, and sort out what I was going through, I thought I could help others dealing with the same challenges and bring hope to this world. Beyond helping myself get better, restore, regain, and now propel myself onwards positively, I wanted and wished the same for other people. Although this motivation was naturally already within me, I did come to realize that as I continually had this thinking to help others, there was more power and strength to propel through what my mind was seeing as difficulties and obstacles. Having this motivation to help others rather than just helping myself seemed to be helping me more efficiently and more quickly get back on my feet and feel more like myself.

One day, my sister and I were talking over the phone, and I shared my motivation with her. She responded that to have such thinking was a sincere, correct motivation that certainly was helping me. She said that my mind and heart were fixating and focusing on greater compassion towards myself and others rather than clinging to a belief of being self-absorbed, self-centered, and selfish due to my current situation. She shared that having such compassion as a motivation would be an active agent that would reveal obstacles more as mirages that weren't really what they seemed.

Although they seemed daunting, impenetrable, impassable, and burdensome at first, these challenges turned out to be mere mirages, appearing intimidating only in my mind. If my mind understood that it could see these difficulties as more unstable, changeable, unfixed, and mutable like a kaleidoscope, a path through, around, over, under, or stepping on would be available to take care of this difficulty or obstacle, and propel ahead. We discussed that obstacles are in the eye of the beholder. How someone with their thoughts and emotions along with their five senses were engaging or the opposite, running from or pretending this wasn't there definitely determined more success or more failure with superseding what was present in their life at this time.

As time has passed, I have contemplated, in retrospect, how I have been propelling through or around challenge after challenge since last March. I found three attributes that have been regularly present. First, that something is larger than me with omnipresence, omniscience, and omnipotence. People could relate to this as some sort of higher power. Others may relate this to an infinite universe or an affinity with an angel or divine feeling, which may be Atma, Buddha, Allah, etc. For me, this is God.

So, there was God around me and connected within me. This presence of feeling was active, alive, and vibrant, not just some thought or belief. Second, the sense of being larger than just my body and my mind. Somehow, I was interconnected to a like-mindedness of abundance, such as people's positive thoughts, which assisted me in developing more powerful abilities and confidence. In recognizing a hidden abundance of like-mindedness, it is essential to recognize something very important about human beings. It is human beings' natural nature as human beings to help each other. As humans, we can't escape this truth even though many people with fear, scarcity, and lack of abundance mindset try to convince us otherwise. Therefore, we should think this, "people around me will and wish to help me. And, I can and do wish to help people." It was my experience time and time again that this occurred to me and to those around me. Third, a continual and endless flow of love. This love felt boundless and constant and was present no matter what. I discovered that loving myself as deeply as possible was paramount to my healing and health.

This is irreplaceable within an abundance mindset. My mother sent me a book with a journal for loving myself that accentuated the depth of love for myself and was a perfect gift at the ideal time. It is not just the love of myself or love from someone; it's the feeling of expansive love that is everywhere, all through everything. This feeling of love is constantly creating and always abundant.

The combined presence of the feeling of God around and within me, the connectedness to like-mindedness, and the uninterrupted flow of love obviously was much larger than any obstacle dancing with me. It was, for sure, much easier to see any challenge more like a mirage that could dissolve rather than a solid barrier preventing any advancement, and I have applied this combination to benefit me in my life. It has steadily helped me progress even in ways people haven't thought possible. To me, this is the complete abundance of health naturally around and within each of us as humans. When this complete abundance is recognized and utilized, there is a great advantage for possibilities and solutions to come forth.

Having all the things that medically happened to me resulted in a lot of "can't." These "you can't do's" came from different directions, including from the medical team, therapists, and well-meaning family and friends. Yes, it was true; I could not do many things and was experiencing these limitations. However, the word "can't" is also an obstacle. There are numerous times that we can actually do something, even partially. However, our mind or someone else's mind has placed the label of "can't" on us or around us, or somehow this label has inserted itself into our mind. Yes, each of us has limitations preventing us from doing something. Even though it may be nice to be able to fly like a bird, our human body is not constructed this way. So, we accept that we have some natural limitations and we cannot do some things. It should be considered that "can't" is often like an obstacle robbing us of our potential, abilities, and victories and is not part of the abundance mindset.

"Can't" is contrary to the abundance mindset. Being interconnected, we are larger than ourselves. Knowing and feeling a higher power within and around us or a fully supportive universe also acknowledges that we are larger than ourselves, so therefore, the word "can't" now becomes diminutive. This word looks smaller and undeniably more insignificant in the mindset of abundance. In fact, being told "you can't" within the energy of abundance now looks like an exciting challenge to prove this is not true. Within abundance, "I can" is active, present, and ready to do.

In reality, many things that I was either told directly or people who just thought I couldn't do unraveled. Why? Because I, as this mind and body, held onto the abundance mindset to propel me through the "can't" and instead see the actual possibility and result. Michael was very instrumental in confronting the "can't" straight on when I was lying on a hospital bed in the ER at Carroll Hospital after the stroke in March. "Matt, Dad would say to people, 'Don't feel sorry for Michael because Michael doesn't feel for himself.' This advice from Dad has remained with me and always helped me. So, Matt, I am not going to feel sorry for you." You are going to get better." It offered a crucial reminder and reassurance that I could heal and get better.

Michael knows about the "can't," so when he said those words about getting through the stroke and getting better, I knew he really meant this. He, too, has his story of "can't," which happened when he was six years old. After a beach vacation in which Michael was limping the entire time, he was diagnosed with Legg-Calve-Perthes, a hip disorder that happens in childhood mostly to boys. The blood flow to the head of the femur bone in the leg that forms part of the hip joint has an interference of blood flow. With little and then no blood flow, this bone dies and stops growing. Healing will occur over time with new blood vessels; however, as the dead bone is removed with new blood flow, there is bone loss and weakening of the head of the femur. Without treatment, the hip joints will sustain permanent damage, resulting in permanent disability. For Michael, both hips were affected.

After about a week in the hospital under traction, for his legs to be separated apart for correct hip joint alignment, he was placed into full right and left leg casts from his feet to his hips with a large bar plastered in between them that set his legs 3-4 feet apart for 9 months. After 9 months, a brace prosthetic in the exact alignment as the leg casts was able to be constructed, which he wore 23 hours a day for another 1 ½ years. He used a pair of crutches to help himself get around. He could not stand for this entire time with the brace off because the hip joints could be irreparably damaged. So, for that special 1-hour break off from the leg brace, he would scoot around on the floor to move. Obviously, everyday life for Michael was a new challenge: sleeping in bed, sitting at the kitchen table, getting into and out of a car, and learning in school, therefore, all of his daily life activities.

My father used to observe how friends, other family members, and even strangers were relating to Michael, feeling sorry for him and, at times, even pitying him. He knew that if this continued, it would be very detrimental to Michael's mind and complicate his ability to deal with this hip disease and adapt to his new living of life. My father wanted Michael to see, know, and feel the "can do" and discard the "can't." He empathically said to Michael, "Michael, I am not going to feel sorry for you, and others shouldn't feel sorry for you either. You are strong, and you will get all better." When he would talk to family, friends, coworkers, or people he would meet with, he would say, "Don't feel sorry for Michael because Michael doesn't feel sorry for himself."

To some people, this may have sounded harsh to say to his child son. My father was saying this out of love and concern for Michael to not feel weak or defeated or so different from other kids that he wouldn't be able to succeed. Michael did a lot of things that people, including doctors, said he wouldn't be able to. "Can't do's" was a word often said to his face. In spite of these "can't do's," he just "did." He figured out how to "walk" up and down stairs in the casts and brace, navigate moving around classrooms and the school hallways, and play with his friends. He really didn't view himself differently from others around him. He balanced himself on the casts or the brace when competing in kickball and used a crutch to hit the ball. Then I, known for my "blue flash" speed, was his runner.

Later, he became a star football player in high school and achieved All-Metro recognition. Michael took this advice from our Dad to heart and, to this day, still says that this advice helped him tremendously and was invaluable. When he advised me the same in that hospital, which was the start of an unbelievable journey, I knew he meant it, and I knew he had lived it with triumph and victory. Now, I would take this advice to help myself propel through the mirages of obstacles, utilizing the hidden abundance within and around each of us that expands our ability and capacity to heal.

Chapter 10: We Are Not a Bunch of Statistics

One of my favorite Bible verses is in the New Testament, James 1:2-4. It says: *"Consider it pure joy, my brothers and sisters, whenever you face trials of many kinds. 3. Because you know the testing of your faith produces perseverance. 4. Let perseverance finish its work so you are mature and complete, not lacking anything."*

If you notice, it doesn't say if you face challenges; it says when you face challenges, it means that all of us will face challenges in our lives. We all have to persevere with whatever we are going through, and we really should not be satisfied with what the statistics say about our disease or circumstance. We might not get through something as quickly as we would like, but we must keep moving forward and not lose faith. We have to keep a clear thought and, even better, a clear vision in our mind that everything will work out no matter how long it takes. Even if you become frustrated, feel unworthy, sad, or even if you think you can't do this anymore, whatever is coming up that is staying "stop," keep going. As we propel through obstacles, we gain wisdom, courage, strength, love, compassion, and gratitude. Rather than losing, we are gaining abundance. This abundance is so complete that there cannot be any lack or need. And, where there is abundance, there is also joy.

Each of us possesses some uniqueness that differentiates us from each other. What may be a faster time for someone may be a slower time for someone else when doing the same task. It doesn't matter. Keep perseverance as your motivator and guide. Don't get tangled up in statistics. I remember hearing the statistical prediction that I would recover from the 1st stroke in March within 6-9 months. On the one hand, it was good to hear that full recovery was possible, but on the other, it seemed like a long time away when I first listened to this prediction. Then the 6-9 months came, and as you can see from reading this book, a lot more happened during that 6-9 month time frame than just the stroke. When this 6-9 month time after the stroke arrived, I wasn't fully better from the 1st stroke, much less everything else that occurred. I did feel very let down and was down on myself until I started to remind myself that everything was going to be alright. I kept saying to myself to have faith that things would be okay and that I was still determinedly persisting.

We, as human beings, are multi-dimensional, such as the multi-facets of a diamond. If we are comparing the facets of a diamond to each of us, the facets are all of our pure positive qualities that are already naturally within us. A diamond is beautiful, can be considered rare, and is precious. Each of us as humans is actually beautiful, rare, and precious regardless of whether we feel or know this or not. Each of us has pure qualities of love, kindness, compassion, generosity, joy, truthfulness, goodness, calmness, peace, and wisdom. Often, people don't speak highly of other people because they don't comprehend themselves as inwardly beautiful, rare, and precious, so, as a result, this cannot be seen in others around them. But I'm here to say people are good because positive, pure qualities are already within each of us, no matter what our past or our current reality. The majority of people are willing to help as long as we are vulnerable enough to accept their help.

Some people could think that somehow, I am being punished because of so many difficulties that have happened with my health. To me, this is a very limited and confused way to think of us as human beings. When a butterfly is about to hatch itself from the cocoon, where it transforms from a caterpillar into a butterfly, the cocoon becomes exceptionally constrictive. The butterfly undergoes an incredible struggle to free itself from the cocoon, and it can look desperate as if the butterfly is making no progress. In fact, it can look like the butterfly is near death. In actuality, the butterfly must push itself through the small opening in the cocoon and strive to escape it because it must push the fluid out of its body and into its wings. If the butterfly does not do this, its body will be swollen, and its wings shriveled, resulting in the butterfly never being able to fly. The butterfly has to face the obstacle to free itself from the cocoon in order to not cripple itself by having shrunken wings. Obstacles and challenges arising within our lives are really not meant to ruin, disable, sabotage or destroy us unless our mind sees them as invaders, punishers and/or demolishers. What I see is that when something is in my way, like an obstacle, it's an opportunity to do something about it. One of the opportunities is to challenge myself and see what I can do to make the obstacle go away.

With all that has happened, can it be said that many hurdles and stumbling blocks were seemingly popping up one after the other? Yes! With all of them present in my life, though, I can share that I have grown as a person, along with my pure positive qualities that are brighter, more powerful, and more abundant. My understanding of my spiritual nature and being connected to God has become more intertwined, and I do not feel separate from God. There's a feeling of a living, vibrant presence ever with me. If I compare myself to the butterfly, I propelled through the obstacles, seeing them more as mirages, which freed me as much as possible from potential impending harm that seemed an actual feasibility. The exertion, effort, and striving around, over, under, or through these obstacles led to an expansion rather than a constriction along with abundance, even in the face of scarcity. As I understood that an expanded opportunity was appearing before me, I was growing within myself and having a much deeper cognizance of love, generosity, caring, compassion, kindness, calmness, peace, and gratefulness. Also, a richness of insights and clarity came into my mind to assist me with more comprehension of my situation, and that maybe the priority of my life would be to help people on a more profound level than I had ever engaged in before.

If we are multi-dimensional human beings who are, in reality, inwardly beautiful, rare, and precious, then how can we possibly be considered as a bunch of statistics defining us and boldly predicting outcomes that can rob us of abundant possibilities and opportunities? Statistics are largely based on research. Statistics are created as a method of measurement to guide a conclusion answer from a preformed question. Some statistics can be helpful, while many others are limiting and only provide a partial story, especially in the field of health and wellbeing. Statistics are reductionistic in nature, only looking at a single sliver of 1 slice of pie and usually neglecting the other slices as well as the whole pie. Approaching life looking at only a single sliver within an entire slice of life occurring at one time that is really within the whole pie of someone's life really looks very ignorant rather than knowledgeable.

In everything that I went through, constantly the "statistics" were being given to me, often with the feeling as if being thrown at me, as if telling me, "This is what you are" or "This is what you will become." Every time this experience happened to me, it very much frustrated me when meeting with my therapists and doctors. I was listening to the statistics of people who had similar health challenges as I did. It usually often felt like they were trying to convince me of the statistical truth, that this is what it is, no matter what. In my mind, I was thinking, "I am an individual and not one of these statistics." "How can they not see that a whole person is sitting in front of them, not just a stroke that happened or a brain mass?"

During a few of the conversations with my sister, I would bring up my disappointment, feelings of anger, and depression, being referred to as a statistic. She fully understood what I was conveying to her. Being in the health profession previously, as well as dealing with our father's renegade cell condition, she had seen this time and time again. She shared with me that the medical and therapy professions are mostly reductionistic, living in a microcosm rather than being able to view the reductionistic microcosm that is inside a broad macrocosm. She was sharing stories of when she was a therapist and ran into reductionistic thinking and statistical references.

When we were talking, she explained she found the same condition with doctors viewing a patient through a narrow, reductionistic lens that often led to debates of the person's potential with therapy that was pushing beyond what the medical approach armed with statistics was saying could be possible. In her view, the field of medicine knows pharmacological or allopathic medicine and not necessarily health. Health professionals can dampen a person's ability to achieve goals if, in their opinion, they say this is not possible based on statistics. My sister shared that she would have done a great disservice to the people she worked with if she saw them as statistics rather than a human being with the potential to tap into.

My sister shared with me that she never liked the word "patient" because it implied an imbalanced relationship of power, like her having more power and her "patient" having disadvantaged power. She preferred to see both of them in a working-together relationship, like a client receiving a service from her and her providing a service to that person. After all, without a person having a brain injury to provide treatment to, even though she had knowledge of therapy to help brain injury recovery, she would have no purpose and no way to use her knowledge. Therefore, a client with a brain injury needed her, and she needed her client with a brain injury to want her knowledge and expertise for support and healing.

These conversations helped me feel better and positively reinforced my thinking, "I am an individual and not a statistic." She would remind me of seeing myself in the totality of a wonderful human being with lots of potential and often referred to me as "Dearest Matt" or "Precious Matt" in an effort for me to counter the programming that I was bound to statistics. It was genuinely important to her that I not hold the idea in my mind that I was some sort of statistic. I do feel statistics have the potential for people to ask, "What is wrong with me?" if they fall outside the acceptable statistic. I find this very damaging and harmful to the abundance mindset of healing and good health. If I had decided that when the 6-9 month stroke recovery time came and went for me, and I still wasn't better, which then led me to just give up and think this was all I could become, where would I be now? I don't think that I would have been encouraging myself to become stronger, better physically and cognitively, putting more effort into making my business more profitable and successful, or increasing my confidence to push out of my comfort zone. Most likely, the idea of writing this book would never have flourished into an actual book. This is why I say to look beyond the statistics and don't become the statistics.

My sister saw the statistic phenomenon with our father when he was in a rehabilitation hospital stay after a difficult hospitalization from the bladder and intestinal obstruction. The trio of chemo, radiation, and the underlying renegade cell situation had driven his white cell, red cell, and platelet counts down. Daily blood tests were being carried out in the hospital to determine if a blood transfusion or an injection to bolster the white cells or platelets were necessary. She watched as the medical community focused intently on these laboratory test results while seemingly almost not seeing or forgetting that in front of them was her beloved father, a precious human being with a living life that encompassed varied living experiences.

The days in the rehab for my father started to revolve around this one number called neutrophils, this 1 specific type of white blood cell with a function to fight infection, this one statistic on whether or not everyone had to wear a mask around him, whether he should have a rehab session only in his room or if he could go into the main gym for sessions and be around other people. What was the possibility he could get infected at any time? Previously, before all this started with the renegade cell situation, my father had been robust, worked out in a gym almost daily, had no history of infections, and was even a regular platelet donor at a local hospital. None of this mattered, though. Seemingly, it was a crucial neutrophil number that was the director of what to do and not to do.

This one number, this one statistic, was beginning to depersonalize my father more and more from others around him. Now, the world seemed to be revolving around 1 number on a lab test, one statistic that was directing his rehab stay days and even his health. My father seemed more withdrawn and depressed because this number literally was a determination of his day. He was becoming more worried and anxious about his health and his strength to get better. Every morning was the question, "What is the neutrophil number today?" It was becoming more and more obsessive, as if my father was being reduced to a single number of a single lab test results.

This was surreal for my sister. She would talk to herself, asking, "is this just me?" She felt she was living in a world of insanity and was trapped in it. This one number, this one statistical measurement of a white blood cell seemed front and center. His body is not this one number. His mind is not this one number. His totality of all that he is can't be defined by this one number.

Fortunately, my father did get home, and of course, my mom and sister dropped the neutrophil thing first because no blood tests were being done at home. Secondly, because he was home and the only priority was for him to feel comfortable being back home, that for sure did not include micromanaging the day based on numbers from a blood test. So, when me and my sister would talk about what happened at doctor visits and the statistics brought up, she would remind me to not get caught up in them and I would focus on the abundance of health that was within me and around me even though things had happened which had impacted my health and ability to do physical and mental activities.

Doctors are humans like the rest of us. They utilize their learned knowledge to their best capacity. Most of the time, their best capacity is viewing conditions or diagnoses through the looking glass of medicine that is defined by medical training, medical literature, medical experience, and medical statistics. This knowledge of modalities and interventions can and often does help people get more balanced and healthier. Conversely, doctors cannot know and will not know everything, while there are things that they don't know, cannot explain, or cannot give answers to. A frustration of mine was that people were put into a statistic when prescribing medication or treatment. Clear accounting didn't appear to be done, considering the possibility that a person may not fit into the statistics for dosing and prescribing medication. I would have preferred that I was being seen as an individual, not a statistic, when medications were being decided for me.

Placing a person into a statistic based on weight and overall appearance is what landed my father into a steroid psychosis and an elevated potassium count, which can cause life-threatening heart arrhythmias and even a heart attack. The prescribing doctor for the steroid said to my mother and father that this dosage would be fine because my father, at this time, was 240 pounds, strong, and muscular. Historically, my father had a lot of misgivings about taking any medications. His same hesitancy was with taking a steroid and Revlimid, a primary chemo drug for Multiple Myeloma. However, he was reassured that he could handle them.

Within a few days of taking the high-dose prescription of the steroid, he was fully paranoid, seeing aliens, seemingly terrified, shaky, and extremely pale. When being medically evaluated, the steroids were completely stopped, and the potassium level went down quickly, with the psychosis resolving within 2-3 days. Not only was my father sensitive to medications, but he had already suffered a massive heart attack resulting in a 5-bypass heart surgery years earlier. Placing him on a high-dose steroid created a huge mental and emotional imbalance with a potential for a heart arrhythmia or, worse, another heart attack. His individuality should have been considered, not the accepted statistic for prescribing.

Doctors are an integral part of the intricate web of maintenance of health and well-being, and this web of care also encompasses a broader range of factors that support, sustain, stabilize, and restore our overall wellness. Doctors contribute to this intricate web with their knowledge and expertise from working with their patients. Being the nature of humans as multi-faceted and multidimensional requires that one must look at our capacity for healing and health through multiple lenses, and these lenses will be unique for each person being their unique human self. One-size-fits-all will not be possible, and why would this be possible when looking at us as humans in a more comprehensive and expansive way?

A kaleidoscope has mirrors that reflect images inside of it, creating symmetrical patterns—the design changes by rotating the section containing the loose fragments. The word kaleidoscope is the Greek words kalos (beautiful), eidos (form), and skopein (view). The form and function of our human body do have beauty if we approach it with a sense of wonder, respect, and appreciation. When we look through a kaleidoscope, we see surprising patterns that we didn't expect or know. These patterns can create within us calmness, joy, relaxation, balance, and a feeling of beauty or a sense of being more aware that we are beyond what we think. The patterns being viewed delicately interconnect themselves within specific alignments.

Looking through the changing lenses of a kaleidoscope can be likened to looking through multiple lenses of what sustains, maintains, continues, and/or restores health. These "lenses for health" are intermingled with each other, and when appropriately aligned, they enhance overall health with more vigor, luster, and vitality. What are these lenses? They are what foster, empower, and bring forth an abundance of health within us and around us. For me, it has been my inter-relational bond to God as a living presence, feeling connected to people's positive thoughts and prayers, doing meditations and prayers, correcting my mind when I start thinking negatively about my situation, continuing to hold the lit match on scarcity to burn it down the ground, embarking on homeopathic remedies, furthering awareness of abundance, actively engaging in personal physical training, and, listening to and partaking in medical and therapy advice while determining what is best for me by asking and listening to myself.

When I was in the hospitals and also in outpatient as well as in follow-up visits, I would at times experience being talked to as if I was a child, incapable of understanding what was going on or what had happened. It was most frequent with therapists, and this child-talk happened much more often to other people receiving rehabilitation rather than to me. Observing this happening to people trying to get better by participating in therapy was very uncomfortable for me. It felt demeaning and lacking respect towards the person who was trying to recover from a health trauma. It's very possible that the therapists were not aware they were speaking in this way. However, this occurring was reducing me or another person from someone who had the capacity to understand to "less than" and "inferior" and being seen as a child rather than an adult.

Feeling diminished, whether intentional or unintentional, does not foster the abundance mindset. It is really in the best interest of our health system to prevent "less than," "inferior to," and "incapable of" towards the ones who are receiving their services. This lack of abundance mindset perpetuated by the healthcare system only takes away while also decreasing the healing potential within a person. This can lead to less than stellar or worse health recovery outcomes, reduced patient satisfaction with health professionals, and discouraged health professionals who can feel undervalued. This is a lose-lose; there is no winner, and this cycle is depleting everyone. A hope of mine is for the health system within our country to embrace an abundance mindset that would benefit both health professionals and the people who are served by their services. A win-win would achieve better results for everyone.

If I were asked to identify an enemy to the abundance of health that expands the capacity to heal, I would have to say the "I have" possession and ownership. Our society has a tendency to label diagnoses as one of these labels. Once given a diagnosis, of course, we will naturally say, "I have this" or "I have that." I realized that if I always was thinking about "I have this brain mass" or "I have/had this stroke," the experience of the brain mass or stroke was going to consume me. It was going to become as if I was possessing this health problem and owning it as being me, like me being this diagnosis and health problem and this diagnosis and health problem being who I am.

Besides the real possibility of making me feel weak and unable to cope with what had happened to me, I was setting myself up to become more reliant on others to help me rather than relying on myself. I was conditioning myself to believe that I was not capable and that I would have to depend on others. I started to catch myself thinking this way and would correct my thinking so I would not fall into this trap. I do believe that doing this really has helped me in a lot of ways, like having more fearlessness to heal, having better health, and doing things that people and initially myself thought may not be possible.

One of the hardest things that can happen to us in our lives is losing self-control over our bodies. This happening to me was incredibly overwhelming. To suddenly have my life turn upside down with so many incapability's due to loss of control of my body was a complete shock. Truth be told, as I was going through all of this, hearing statistics was not what I needed to listen to at that time. Loren was also sitting next to me, listening to all the statistics and trying to absorb what it all meant for me, her, our family, and our lives. Statistics can be part of a slice within an entire pie, and it would be more helpful to handle statistics with care for people who are in the maze of recovery after experiencing health difficulties.

Working through the maze of what helps as a path to better health and what is a dead end requires flexibility of analysis, thought, and application into a viable health recovery and health plan. As a part of a piece of the pie, statistics should be recognized just as this and not as a prime indicator of a person's potential health outcome. There's a hidden stress and anxiety I think many people may have gone through when being given these statistics around healing and health is the question, "Am I going to be within the statistic of successfully getting past this health challenge, or am I going to be within the statistic of failure and not getting past this health problem?"

I know from my own personal experience, as well as being in my business, that people need to have something to work towards, especially when some sort of challenge or obstacle is preventing achieving success. People also have to have trust in who they are working with to dissolve or move out of the way, whatever it is, so success can be achieved. If a person is in the midst of an obstacle that is impairing good health, he or she needs to actively contribute towards restoring, sustaining, and maintaining a good level of health. Passively relying on or expecting someone else to do this for himself or herself reduces the power of the abundance mindset. Therefore, taking health seriously, respecting health, and looking at things differently in your life is imperative. Directing your abundance healing power towards changing the way your life is being lived as well as to what is interfering in your life that is jamming up or shutting off good health has to happen to give you the best possible abundance of health and healing.

I have experienced great success in my life, and now I have been dealing with the biggest challenges of my life. All of us want to have the comfort of knowing what's coming next in our lives, and all of us can experience anxiety when we don't know what that may be. Since we like to know what we think is coming next in our lives, we are over-scheduled and over-planned, which, as a result, leads us to forget about the beautiful parts of our lives that are already occurring and what our life has to offer. Within all the experiences that have been presented to me, this has been the time for me to reflect and accept that marvelous things have happened along the way. Taking time to appreciate and let sink in the beauty that is currently occurring in our life, along with pacing ourselves to restore and recover with love rather than thwarting ourselves with frustration, is a necessary ingredient to prosper the abundance mindset.

From the business that I do, as well as what I see within my own life, being consistent is very important. The point I'm trying to make is that when you are consistent with everything that you do, you are more focused, and there is a less of a chance that distraction will take you off course. Being a consistent person to yourself, a consistent friend, a consistent spouse, a consistent father or mother, a consistent colleague, etc., allows you not to get caught up in what the rest of the world is doing. You just stay consistent, and you will be successful no matter how long it takes. I have been and continue to use this formula as a part of the abundance mindset to increase my healing and health, and I have positively benefitted.

One thing that I think should be brought up regarding us as humans and how we can best heal is to look at the negativity that is constantly happening all around us. I'll never forget a talk I went to several years ago. It was a talk about happiness and the fact that happiness is not a mindset but a skill set within our country, the United States. Here we are as the most prosperous nation in the world, yet we're also the most depressed nation in the world. Young people are becoming depressed at younger and younger ages each year. We are surrounded by negative news 24 hours a day, seven days a week. With the advent of social media, many young people who are not happy, don't have a safe space to go, that gets them away from whatever's making them unhappy.

People are addicted to negativity, not realizing they are addicted because negativity has become so commonplace, and has an expectation that it will be present no matter what. When I was in the proton center waiting area before radiation, there was a TV. The TV had on a popular series about a murder that had occurred and the special unit that was working to solve the murder. Here was a physically beautiful building, obviously constructed to bring in natural daylight and help create a more uplifting atmosphere. The effort of this was lost while I was in the waiting area, watching what was on the TV and what the script of the show was about. While I was working hard to block out the noise from the TV, other people there were seemingly glued to watching the TV show. All I wanted to do was focus on being calm and thinking good thoughts.

I don't think these people realized the effect on their minds and bodies by watching the negative theme of the show, along with the sights and the sounds. They already had a very serious disease that was literally harming them. Then why should they feed their mind with a TV show focused on murder and solving a murder before having a radiation session? When my father had to go to the infusion lab for chemo, he often came back with stories of people eating various junk foods while the chemo was being pumped into their veins. Watching or listening to negative topics is like feeding the mind a negative diet. As we know, junk food feeds the body a negatively nutritious diet. Negativity can't feed an abundant mindset, and life will not thrive or prosper in the face of negativity. Negativity's hand can only offer poison. Abundance's hand can provide nectar for good health.

Negativity is so prevalent in our lives that it seems to be the usual and normal way of living. It can get to the point where it's practically impossible to think or be positive. Very often, many people are consumed by what is going on in the world and are distracted from making their individual lives more abundant, fruitful, and satisfying. Years ago, I read a book that talked about the power of negativity in our lives. In this book, there was a reference to being in a reasonably positive household. Even in this reasonably positive household, exposure to negativity through such things as speaking to each other, listening to TV or other media, reading the newspaper, being in school, being at work, and going out to do things like shopping or to a restaurant was staggering. What was most poignant was how much negativity a child had absorbed at a young age. This book really made me think, "What is going on with us?" People don't realize the effect of negativity on their thoughts, emotions, and bodies until they start to cut back from being exposed to it. The more I cut out either listening to or being a part of negativity, the more I realized that life is so much better. I am calmer, more at peace, and can enjoy living more. Without a doubt, it is easier for me to feel happier and more secure. And I know that less negativity in my life has positively affected all areas of my life, including my capacity to heal, and I have much-improved health and overall well-being.

As a business trainer, when first meeting with a manager or business owner, I usually ask two questions, "Tell me what your sales team is doing really well." The 2nd question is, "What, if anything, would you like to change?" Why do I ask the opening questions this way? As for me, it is more important to know what is going right first and if there is a feeling for any necessity of change in what is already being done. I want to know what is the positive or what is going right. I choose not to start by asking what the problem is or what is going wrong. If anything is going wrong, it will be recognized at the proper timing, with an action plan in place to shift or correct it to a positive direction and outcome.

Many times in our society, we start with what is wrong, what is not working, what is not right. Our mind is already feeling stuck, confined, helpless, powerless, or frustrated. What is more alright, more positive or somewhat working doesn't have focus and is negated. Working through a health challenge from this negative perspective already puts us at the very bottom of a huge hill to climb rather than maybe possibly being halfway up if we were looking at what is alright, what is working, and what is positive. Mostly, we are caught up in the negativity of our health situation, what's going on around us, and even probably what's happening in the world. An abundance of health mindset would say, don't get caught up in the minutiae of statistics, don't get caught up in the negative thoughts and emotions you and others around you are having, don't get caught up worrying or being upset about what's being reported that's going on in the world. Instead, focus, persevere, be aware of directing your abundant healing power beneficially, see obstacles as mirages to propel through, keep reminding yourself to have faith that things are turning for the better, and most essentially, love precious you as deeply as possible.

Each of us, as human beings, has a choice of our mindset. We can either have a mindset of scarcity or lack or a mindset of abundance and possibility. This leads me back to James 1:4-5 4: *"And let patience have its perfect work that you may be perfect and entire, lacking in nothing 5: But any of you who lack in wisdom, let him/her ask God, who gives to all liberally, and it is given to him/her."* This verse gives me incentive and reminds me to cultivate and maintain an abundant mindset. Contrary to what we may think, patience, one of our inner wisdoms, is very helpful for us with our abundance mindset. When we are more patient, we can "see the forest through the trees" rather than reacting impulsively or irrationally. In other words, deeper meaning and understanding of what is happening as well as why can be exposed and seen more clearly. When we take this information with acceptance, we often grow gratitude for this knowledge that helps us proceed through difficulties with more ease towards a more perfect favorable outcome. As part of the abundance mindset, when you feel impatient, delve confidently within yourself, breathing into and out of yourself. Let yourself become aware that something is larger than just yourself; you are interconnected through perfect, abundant positive qualities within and around you, and an endless flow of love is within you and around you. Surround yourself with people who have similar abundance mindset so you can continue with your abundance mindset consistently.

After reading about the power of negativity, you may be thinking, "Yes, it's truly possible that my mind is consumed by negativity more than I realize. I don't really know how to start to change, stop, or get out of this negative thinking maze." The answer to this quandary is to facilitate your hidden abundance as a corrective countermeasure to the negativity.

When we were either talking over the phone or on FaceTime, my sister and I worked together with specific meditation and energetic methods. I am sharing one specific energetic meditation method we would do together. This method fosters, stimulates, and strengthens your more direct awareness of your hidden abundance. As your hidden abundance reveals more, you will have the ability to direct it toward healing and health, including your well-being. The hidden abundance method incorporates the 3 attributes of something larger than yourself, interconnectedness and a flow of love. It is very helpful to do this method daily and, especially if you are facing any crisis of any kind, as it can create a change of thinking, direction, and energy in your mind that can help foster more calmness, peace, understanding, acceptance, and possibly usher forth a solution to a problem.

This method is called:

3 Lights of Continuing Abundances—Accessing Hidden Abundance

1. Sit or lay easily and quietly, with a relaxed body position; close your eyes
2. Take 2-3 deeper breaths into yourself and breathe out to bring yourself more present to where you are and slow thoughts down.
3. Take 3 breaths into your chest at your own pace to open some space within your chest.
4. Take 3 breaths about 5 fingers above your belly button to open some space within your abdomen.

5. Take 3 breaths in and out of your nose into and out of the center of your brain to open some space in your head.

Now, the 3 lights of continuing abundance:

1st light: White

1. From the center of your chest, think about or see or feel bright WHITE light going out from your chest all around you and as far out as possible, thinking it is going to all the world and beyond into the sky, into space, and beyond space.
2. This light is love, kindness, and compassion. Repeat, "Love, Kindness, Compassion," as the WHITE light is leaving you and going everywhere. Then think that this WHITE light is transforming everything around you to **Love, Kindness and Compassion, Peach Joy**
3. When the light gets to its furthest place in your mind, then think, "Here is all the abundance of Love, Light, and Compassion, Peace, Joy"

Now, gather all this abundance of Love, Light, and Compassion, and let the WHITE light filled with this return to the center of your chest and dissolve within you completely filling you with Love, Kindness, and Compassion.

2nd light: Yellow

1. From the center of your chest, think about or see or feel bright YELLOW light going out from your chest all around you and as far out as possible, thinking it is going to all the world and beyond into the sky, into space, and beyond space.

2. This light is Perfect Wisdom and Perfect Knowledge. Repeat "Perfect Wisdom and Perfect Knowledge" as the YELLOW light is leaving you and going everywhere.
3. Then think that this YELLOW is transforming everything around you to **Perfect Wisdom and Perfect Knowledge.**
4. When the light gets to its furthest place in your mind, then think, "Here is all the abundance of Perfect Wisdom and Perfect Knowledge."

Now, gather all this abundance of Perfect Wisdom and Perfect Knowledge and let the YELLOW light filled with this return to the center of your chest and dissolve within you completely filling you with Perfect Wisdom and Perfect Knowledge.

3rd light: Dark Blue

1. From the center of your chest, think about or see or feel DARK BLUE light going out from your chest all around you and as far out as possible, thinking it is going to all the world and beyond into the sky, space, and beyond space.
2. This light is Protection From All Harm. Repeat, "Protection from all harms," as the DARK BLUE light is leaving you and going everywhere. Then think that this DARK BLUE is transforming everything around you to **Protection from all harms.**
3. When the light gets to its furthest place in your mind, then think, "Here is all the abundance of Protection from all harms."

Now, gather all this abundance of Protection from all harm and let the DARK BLUE light filled with this return to the center of your chest and dissolve within you completely filling you with Protection from all harm.

Just let yourself be aware of as you rest in abundance: Eyes closed

1. Abundance of love, kindness, and compassion is within you and all around you
2. An abundance of perfect wisdom and knowledge is within you and all around you
3. Abundance of protection from harm is within you and all around you

When you open your eyes, reaffirm to yourself:

1. "Abundance of love, kindness, and compassion is within me and all around me."
2. "Abundance of perfect wisdom and knowledge is within me and all around me."
3. "Abundance of protection from harm within me and all around me."

When ready, get up and move around with confidence that these abundances are yours.

This method could help you with your hidden abundance. Practicing the method regularly, just like we start an exercise program in a gym or a swimming program or running program, is important and paves the way for the prosperity of your hidden abundance. As I shared earlier, consistency is the key to living a more successful life and living more consistently in your hidden abundance will surely benefit you in numerous ways. So, therefore, consistently applying the 3 lights of continuing abundance helps bring about positive change while greatly diminishing and decreasing negativity.

As this book is coming to an end, you may be wondering where I am now from, traveling this very unexpected road that started after the first stroke. I had made that a goal to get back to the gym with Michael. We go to the gym together 2 days a week, and he continues to push me to gain more strength, coordination, endurance, and integration of my body and brain working together. Week by week, there's more progress and I am gaining more good physical condition. I look at these changes regularly, and it really feels great to see where I am now from where I was in the past after finishing radiation. My business life continues with the franchise that I merged with, new clients have come on board, and the opportunity for business growth continues to expand.

To my pleasure, I'm getting back into taking care of things inside and around our house, like yard work. Life with my family continues onwards with positive changes, such as Lucy graduating from high school and getting ready for college. I couldn't be prouder of her than I already am. I am thankful for all my children and Loren and very proud of each one of them. I am more comfortable and able to be around family, friends, business colleagues, and meeting people who I don't know without being overwhelmed or feeling insecure. I love myself more deeply than I have ever before, and I love everyone in all my family so much. They all mean the world to me. My emotional and spiritual heart has prospered beyond before everything began. My mind continues with the sincere wish that each person in our world becomes happier and that everyone has good physical, mental, emotional, and spiritual health.

In closing this book, whether it's you with a health problem or you are a family member or friend or another, I very much am hopeful that reading this book has been helpful for you in some manner or fashion, no matter how small or how large. I can only share my story as it has worked for me with the wish that somehow it enhances or assists your capacity within yourself to experience the abundance that grows and expands your healing and good health. I am reminding you to be sure to celebrate the smallest of accomplishments as true victories and triumphs that are sprinkled with gratitude for yourself and the others who have helped you. I am enormously appreciative with a feeling of pride of where I am now and where I have come from when the unexpected shake-ups started and proceeded outside of my direct control. I have this same wish for you: that you will be extremely appreciative and proud of yourself wherever you are on the path of healing and health and that you will be able to realize where you have come from with a feeling of some wonderfulness for you and your life.

Acknowledgments

The idea to write this book was just that: a daunting thought seemingly sat in the air. This thought that originally germinated as an idea has seeded and now bloomed into fruition. Filled with gratitude that my life is still present on this earth, I also share immense gratitude to the literal "village" that supported, stood beside me, encouraged me, and believed in me to write and finish the story of this book.

To Loren, my long-term partner, my lovely wife, my ardent caretaker, and from my heart as your devoted husband, thank you for being you, especially during this unprecedented time in our lives that could have torn us and maybe our family apart. Even under such strain, you continued onwards with resolve and strength of will to plan, coordinate, and assist with my recovery and progression to better health so I could write this book.

To my loving mother, you held me close in your heart as your son, constantly sending me your protective motherly energy to me that I would be protected from all the health challenges that were happening, become better, have good health, and have the best yet to come in life. You never gave up on me, and you have supported me in many ways that have encouraged me to have the confidence to write this book.

To my lovingly departed father, you had a good heart and always had my best interest in your heart. You certainly faced and determinately accepted pushing through adversities in your life that showed me to keep going even when the going got tough. "Atta boy, Dad" for displaying courage within adversity that has been an underlying component in writing this book.

To my close sister, what we have shared together these past few years propelling through these health challenges. It is so very much that a few words cannot express the depth. You have been devotedly involved with every step, from concept writing to publishing this book.

To my children: Molly, Georgia, Jack, and Lucy, each of you is and will always remain a source of inspiration and motivation to write this book. Molly, Georgia, and Jack, your selfless willingness to become a kidney donor for me is beyond admirable and has no sufficient words of gratitude. Lucy, you creatively cheered me on and that helped me get through all these health difficulties.

To my special brother, Michael, without you, I would not be where I am today from what has occurred since March 2023. You stood by me as a pillar of steadfastness and sturdiness, constantly reminding me not to quit or give up right from the first day when this unexpected, very bumpy road started. "Don't feel sorry for yourself" was extremely important to hear and be reminded of, which fostered increased strength within me. Your power of consistently seeing potential within me has spurred me on to write this book.

To Loren's parents, George and Kathy, you deeply helped Loren with strength and guidance while supporting us and keeping the house going at a very pivotal time in our lives. This support helped me to restore and also gave some space to write this book.

To Bradley and his family, for their kindness and love towards Molly, me, and our family. For Bradley's foresight to start a GoFundMe page and all of those who generously donated helped buffer the debilitating financial strain and also helped to support writing this book.

To the hospitals, doctors, nurses, therapists, other medical staff, and support staff that provided medical care during these health adventures, your dedication, and skills helped me with the ability to hold onto my health and push myself forward towards restoration to usher better health that has allowed me to write this book.

To the business franchise that offered me the opportunity to become part of their franchise as a solution to my business dilemma, the opportunity to expand into a newer business endeavor has helped financial sustenance that has provided a way for me to heal more fully and also have a chance to write this book.

To my extended family, my friends, my family's friends, my neighbors, my church congregation, and my business associates, you all helped support me in many ways, for which I am thankful, and has been a contribution towards me writing this book.

To all the people who I know and those I do not know who came to know about my situation and sent positive thoughts and/or prayers. All of this assisted me with endurance to be forward-looking step by step as I became better, which has helped me write this book.

To the publisher, who asked, "You faced a health adversity? Was there a victory?" to which I replied, "Yes." The response: "People, especially at this time in the world where there seems to be so much negativity, need stories of triumph coming out of challenges."

To the audience of people who will read this book, without our commonly shared or related experiences, there would be no need to write this book. My desire to help others with what I have been through and shine a ray of hope is a deep motivation to write this book.

To my chosen spiritual path of following Jesus as my Lord and Savior and knowing the presence of God. The message of abundance of physical, mental, emotional health, and spiritual health is the point of writing this book.